Unspeakable

'I was gripped by *Unspeakable*. These patient stories take the reader right into the consulting room and show how psychotherapy has the potential to transform the lives of those who have encountered unimaginable trauma'. Caroline Elton, author of *Also Human*

'A humane exploration of traumatic experience and, more importantly, survival, that resists simplistic formulations and captures the subtle drama of psychotherapy.' Frank Tallis, author of *The Act of Living*

'With intelligence and deep empathy, Dr Gwen Adshead and Eileen Horne bring voice to acts, experiences and emotions that defy language. This is a book that challenges assumptions, expands compassion and lingers with you long after the final page. Fearless, humane and beautifully written – *Unspeakable* asks what happens when we finally face the stories we've spent our lives avoiding.' Emma Reed Turrell, author of *What Am I Missing?*

'Wonderful . . . Adshead shows how the mind is infinitely adaptable . . . We learn that it's possible not only to survive traumatic experiences, but to find new ways to flourish in spite of them.' Gavin Francis, author of *Recovery by the same authors*

THE DEVIL YOU KNOW

Dr Gwen Adshead & Eileen Horne

Unspeakable

Stories of Survival and Transformation After Trauma

faber

First published in 2026
by Faber & Faber Limited
The Bindery, 51 Hatton Garden
London EC1N 8HN

Typeset by Ian Bahrami
Printed and bound by CPI Group (UK) Ltd, Croydon, CR0 4YY

A CIP record for this book
is available from the British Library

ISBN 978–0–571–38524–9

Printed and bound in the UK on FSC® certified paper in line with our continuing
commitment to ethical business practices, sustainability and the environment.
For further information see faber.co.uk/environmental-policy

Our authorised representative in the EU for product safety is
Easy Access System Europe, Mustamäe tee 50, 10621 Tallinn, Estonia
gpsr.requests@easproject.com

2 4 6 8 10 9 7 5 3 1

To all of those brave people who are
at the beginning of the road

The past influences everything and dictates nothing.
ADAM PHILLIPS, *Darwin's Worms*

Contents

8 The Trainee

x

Introduction

This is not a book about trauma.

It is a book about survival – especially the experience and quality of survival that may be possible after suffering psychological wounds that damage identity and take away language. The stories contained here are based on personal encounters that I have had during three decades of working with people who are struggling after deeply distressing and often life-altering events.

After qualifying in forensic psychiatry, in the early 1990s I started work as a psychiatrist and researcher at one of the country's first trauma clinics, in London's Maudsley Hospital. The clinic had been set up following a series of mass disasters in the UK; previously, the focus of most funding for trauma care and study was on soldiers, as it has been since ancient times.

Trauma, derived from the Greek for 'wound', was once generally understood as something that needed medical treatment. Today, it is a word in such common parlance that it risks losing all meaning. Is it a 'trauma' to fail an exam or have a bad break-up? Or are these things very sad but ordinary upsets? How do they compare to the experiences of being a soldier in wartime, a victim of violent crime or a refugee? Who gets to say what counts or which kind of trauma is worse than any other? Certainly not me.

As I began to write this book, the director general of the World Health Organization made an extraordinary public statement, saying 'the whole world' was suffering from 'mass trauma' in the wake of the COVID-19 pandemic.[1] Perhaps this response was understandable, given the fear the pandemic caused globally and the massive fallout, both medical and societal. But I wonder if there is not some risk in suggesting that any large group of people, let alone 'the whole world', all share the same psychological response. There's an implication that prior to 2020, we were all the same, with similar childhoods, community experiences and support systems. This cannot be accurate. Even if a group of people experience the same awful event, as some of the following stories describe, everyone's response is complex and unique to their circumstances.

In the years that I have been working as a psychotherapist, it is safe to say that each person I have seen has been through something that could be called 'traumatic'. The causes of trauma can vary widely, but it usually stems from an experience that is startling, frightening and disturbing. This activates the nervous system's adrenaline response, an energy surge that enables people to run away, fight or freeze in the face of threat. The hormone cortisol then kicks in to release glucose, giving the body energy for a longer-term stress response, if necessary. All of this is well known and well worn in the public discourse; 'fight or flight' is as familiar a phrase today as the word 'trauma' itself. Unfortunately, generalising about how our bodies respond to threat can mask a welter of psychological nuances and complexities. Trauma and our responses to it are not just one thing. There are many colours and gradations of trauma, just as there are many versions of survival. For instance, brief one-off events usually have less impact than longer, more chronic experiences of trauma. Social context also matters, especially when trauma is caused by other people, where the relationship with the perpetrator can be significant. After a single traumatic experience (like a road-traffic accident) most people feel stunned, numb or detached for a few days or even a week; then, over the following weeks, a post-trauma response is likely to develop. Most commonly, this involves some anxiety and depression, but also the well-known symptoms of hypervigilance and re-experiencing phenomena (such as nightmares and flashbacks). Not everyone has all of these, and a small group of people have no such responses at all. Typically, the acute post-trauma response resolves spontaneously over the next six to twelve months. If that doesn't happen, the consensus in the mental health field is that the person has post-traumatic stress disorder (PTSD), which can become a chronic condition.

However, the diagnosis gets complicated in cases where the trauma has been recurring and sustained. As I will describe, it is also dependent on the age of the person when the trauma happens and what other distress they may have experienced in their life, both before and after the event in question. It is one of the injustices of human evolution that the more vulnerable to stress you are, the worse your outcome. It is also a reality that life does not stand still while people are recovering from trauma, and their PTSD is likely to get worse and persist longer if they have to face new, additional stressors.

My clinical experience in secure hospitals and prisons has taught me that violence perpetrators are often also trauma victims, and this understanding has given me a deep appreciation of the complexity of experiences that are too terrible to speak of and that change identity forever. I've seen how painful it is, even (or especially) for someone who is the survivor of a disaster where, as the perpetrator, they *are* the disaster. People feel indelibly marked by a guilty verdict, and I've seen how much time is needed for their narratives about themselves to shift. I've written about the disbelief that some offenders have about their actions, a feeling that 'I am not that person'.[2] I have learned that I need to listen to what they have to say about what they have suffered, as well as what they have done, no matter how hard it is to hear. It is attentive listening that helps people change their minds for good. This perspective proved invaluable as I moved outside the high walls of secure institutions and prisons to work with non-offenders – people seeking treatment in NHS trauma clinics or needing assessment for trauma-related legal reasons. I draw on that work in the following stories, including many encounters with people who, unlike my forensic patients, had never seen a mental health professional before and, in some cases, were not even sure they needed one.

While there is no manual for how to live after trauma, language – and the capacity to think new thoughts about oneself – is a vital part of recovery. As Cicero had it, language distinguishes us from other animals; speech is inseparable from human growth and identity. The process of therapy is about helping people find words to 'speak their minds', but with trauma survivors, it is common to struggle to do that, sometimes to the point where they may only be able to sit with me in silence. As part of my training to become a psychotherapist, I had to have therapy myself, and I remember my own speechlessness. Sometimes I was struck dumb when I remembered and revisited painful stories from my life, trying to find a way to articulate them. When I switched chairs, this made me better able to sit with my patients and keep them company until the right words came – once they were ready. The readiness is all.

There are several myths that have developed in recent years about life after trauma. I hope the encounters I describe will challenge some of them. One is that traumatic memories are permanently embedded in the brain, an asbestos that permeates and poisons unless it is dug out. The best current research suggests that memories are not like this at all; they are malleable, fragile and fluid; they change and may even disappear with time, place and conversation. Sometimes they are protective, too, and help shield us from further harm. Clinical approaches to remembered pain can occasionally focus unhelpfully on the wrong that's been done, rather than on the destructive feelings that may arise, stopping us from moving forward. This may be as unproductive as trying, in the forensic realm, to make violent offenders feel bad about what they've done, instead of seeking to understand how their life experiences may have increased the risk of them inflicting cruelty and harm, and how to reduce that risk in future.

Recovery after trauma does not always require the 'rooting out' of memories. Unpacking every detail of a painful and devastating experience could actually cause more damage, especially if done at the wrong time and in the wrong way. Memories are constructed anew when the mind has need of them, and they become, as the philosopher John Locke first suggested more than three hundred years ago,[3] a crucial part of a person's sense of self and identity. The therapist tries, as best they can, not to interfere with that process. They let the speaker lead and go where they choose, because to do anything else is to take away agency and choice, which would only mirror many people's painful experiences of trauma.

Another belief that I'd like to explore and question is that trauma causes irreversible damage. This ignores the fact that across our lifespan, our brains and minds are plastic (or, as I think of it, *dynamic*), especially in childhood. Yes, people are changed by traumatic experiences, but the nature of that change may be different at different stages of life. In addition, post-traumatic growth is as possible as post-traumatic stress, and in my experience, it may be more common.

Hemingway said that 'The world breaks everyone',[4] and we see this idea widely reflected today, but how is it that some people are so much more resilient than others? In my forensic work, I've adopted the useful metaphor of the combination bicycle lock to explain how risk factors for violence must line up to release the cruelty of which we are all innately capable. The tiny minority of the population who harm other people have a fateful combination of 'numbers', ranging from things like poverty and isolation to substance misuse to simply being young and male. There is always one final, highly personal 'number' that must click into place to release the violent action. So, too, with trauma survivors: in every one of these stories, you will discover that there are protective factors that can align to help insulate people from the effects of their trauma, and other factors that, in combination, may predict a lower quality of survival. The final, determinative 'number' is often related to their earliest attachment relationships with primary carers, which lay the groundwork for later resilience or vulnerability.

I accept that trauma experiences can shatter past identities. A married woman is suddenly and violently widowed; a child is removed from the family and put into an abusive care home; a man is mugged, robbed of his wallet and agency: in these circumstances, survival involves a reconstruction – a transformation of identity, heart and mind as people build a new sense of who they are now.

There has long been debate about the language of the trauma experience, especially the use of terms like 'victim' versus 'survivor'– not so much in the mental health field, but on social media and in our culture more generally. I tend to favour the latter, because after all, if you are alive after trauma, you have survived. The real question is, at what cost? And who do you want to be now? Some people need to hold fast to the mantle of victimhood, whereas for others, being a survivor – with the agency that the term implies – is vital. Similarly, some find mental health diagnoses helpful, whereas some resist them. A number will find relief in being compensated for harms endured, but in my experience, adversarial legal processes can also complicate recovery after trauma. Financial recompense can bring a sense of justice, but this is not the same as healing.

I am passionately interested in words and how we generate the stories of our lives. In this book, I focus on people whose capacity to speak their minds was impaired in different ways. They may have chosen silence, or shock may have muted them; some were threatened into silence by others. Their language may also have been affected by shame, that worm-like, destructive emotion that burrows deep into the soul and multiplies, unchecked, leading to isolation and loneliness.

Some of the silences I describe in these pages are explained by external factors as well as internal emotions. In the wake of trauma, people may be choked by not only rage at what has happened, but ideas of how they 'ought to' respond. I fear there is a stereotype of what a respectable, acceptable trauma narrative sounds like, which can hamper those who need to express strong feelings that don't fit into that framing. This seems especially true for women and those who feel othered and vulnerable. Challenging this norm may be a matter of urgency, if we think of the communities of children who are being traumatised across the world right now. If nobody helps them to speak, what will they do with their anger?

Most accounts of trauma in our culture seem to accentuate helplessness and neediness over feelings of anger or the wish to get revenge. It is perpetrators who are generally caricatured as angry monsters, yet I believe I've seen more rageful and hostile people for trauma therapy in the community than I've ever encountered among violent offenders in secure institutions or prisons. You will meet a few of them here.

Working with my friend and co-author Eileen, we identified some recurring questions about trauma that I'd heard a lot from students, patients, fellow professionals, journalists and others. We then drew from case stories that had made me think differently and might offer some answers or stimulate new discussions. These involve men and women of various ages and backgrounds, covering a range of traumas that encompass everything from single incidents to chronic abuse suffered at different stages of life, at the hands of either strangers or close relations. In each one, we explore those 'bicycle lock' factors that can have a significant bearing on people's responses, as well as on their treatment. Many of these stories draw on my years working in two London trauma clinics, first at the Maudsley and later at the Middlesex Hospital. We have also made use of assessments I've done, both in medico-legal cases and as part of my general psychological work in the community.

For those who would like to go deeper, the end notes list some academic studies and books that I have found helpful. The trauma literature is vast, whether written by experts, professionals or survivors. This book is simply my small corner of that enormous tapestry, woven with my memories of the stories I have heard and the lessons they have taught me. Its purpose is to illuminate the reality of these human lives and offer hope.

Dr Gwen Adshead
August 2025

Authors' Note

In creating this book, we have drawn directly from trauma experiences described by real people. Each story is clinically accurate, but we use composites and have changed or fictionalised names, places, professions and other details to ensure we protect everyone's essential right to confidentiality and privacy.

Dr Gwen Adshead and Eileen Horne
August 2025

The POW

'Mr B? Thomas B?'

Standing at the door of my consulting room, I scanned the large waiting area and saw two men rise from their seats. The younger man stooped to grab an umbrella from under his chair, while the older one carefully folded a newspaper – *The Times* – and glanced over in my direction when I repeated his name. This was my patient. 'Wary' and 'precise' were the first adjectives that came to my mind. Something about his close-cropped white hair and posture suggested a former athlete, perhaps a swimmer or gymnast, poise as well as strength.

I wasn't expecting two people, and reached for another chair to bring into the room, then watched as the younger man, of taller, broader build, with a stubbly beard and an unruly mass of dark hair flecked with grey, took the older man's elbow, guiding him a little as they passed the others who were waiting to be seen. I thought there was care and concern there. A big age difference, too; could they be father and son? Or uncle and nephew? However, both their physical dissimilarity and something about their interaction made me think they were a couple.

I had not been working at the clinic for long, and I noticed that my new colleagues called out people's names and then walked over to escort them, especially on first meeting. During my training in other hospitals and institutions, I had preferred to let people come to me, valuing those initial moments of observation as a new patient approaches. I could learn something from seeing them at a little distance before contact. Just as a detective stands and observes a crime scene or studies the face of a suspect, so I was trying to be alert to the smallest detail at that initial meeting. Sleuths and psychiatrists have much in common, not least because our work often begins in the context of trauma of some kind. In fact, as medical students, we were explicitly encouraged to think of ourselves as 'gathering clues' to develop hypotheses about our cases. The task became more challenging when I moved on to specialise in psychiatry. I remember one of my first psychiatric clinics as a trainee, when I saw a patient who had suddenly developed acute depression and anxiety, for no apparent reason. We were stumped, until his wife mentioned that his handwriting had become cramped and small over the previous year. I realised she was describing micrographia, which can be associated with a brain tumour. How pleased I was to have caught this. I ordered a scan, which may have saved his life. But that experience also taught me that I could just as easily have failed to diagnose the tumour if I had not been paying attention or not asked the right questions.

Mr B's GP had referred him due to an equally sudden onset of disturbing symptoms. In the early 1990s, there was still a convention of old-fashioned courtesies in the medical referral process. A letter would come, addressed to one of the senior consultants or the professor who ran the clinic, asking them to 'kindly see this charming gentleman who suffers from . . .' – whatever it happened to be. The letter was now open on my desk, the details scant, as the GP described 'this delightful 72-year-old who is generally fit and well with no psychiatric history. He has complained to me of poor sleep and recurring nightmares, which appear to be connected to a time he spent as a prisoner of war fifty years ago. He tells me that he has never been troubled by them in the past. I would be grateful for your advice about appropriate treatment and management.'

While everyone who came to the clinic had a unique story to tell, they often shared similar symptoms, with nightmares featuring high on that list. Although we don't know precisely what causes them, they seem to be evidence that memories of trauma are still live and unprocessed in the mind. We often saw people with a delayed response to trauma, too. They would manage to block out distressing experiences for days, weeks or months, until something released them into consciousness, exploding like a depth charge. I was amazed that the referral suggested this man had not experienced symptoms for fifty years. How was it possible for trauma to lie dormant that long?

The trauma clinic at the Maudsley shared space with other hospital services, so there was an examination couch and a washbasin in the room, with my small desk facing a chair for the patient. I have never liked to see people across a desk, so I had arranged two chairs at forty-five degrees to each other. Now I quickly dragged the third chair to create a semi-circle. I smiled, welcoming the two men in and asking them to make themselves comfortable – a common platitude so much easier said than done, especially in this setting. Hands in their laps, the two men sat gazing at me expectantly, strangers in a strange land.

Most of the people I'd worked with in forensic services had long histories of mental health issues and were all too familiar with psychiatrists and their questions. Here, my patients had often never seen a mental health professional before and had no idea what to expect. I think they feared some sort of impossibly difficult interrogation that would end with their detention in a gloomy institution. It could be dispiriting and even bleakly comic to see how uneasy people were, and alleviating that was my first task.

I introduced myself and addressed Mr B by his full name. 'Please call me Tom,' he responded warmly, and I noted how he was making good eye contact with me; no guarding or suspicion there, although I saw his leg was moving with that unconscious knee joggle that can denote excess energy or anxiety. I looked at the younger man, leaving a pause in which I hoped he would say who he was. 'I'm Albie,' he said. 'I'm Tom's friend . . . come to give him support.' Was there a minuscule pause before the word 'friend'? At that time, a same-sex partner did not have the same status in a medical context as the next of kin, which might have concerned him.[1]

Albie's presence needed a bit of thinking about. Most people bring someone to their first medical appointment for a mysterious and worrying problem, and psychiatry is no exception. I wanted to make sure (thinking back to the wife who had told me about her husband's handwriting) that I wasn't missing out on some vital information from Albie about Tom and his life. But good doctors also know that even the most charming partners can be a source of stress or even a cause of trauma. Without a one-to-one conversation I couldn't know if Albie was part of the problem.

I began by telling Albie that it is always good to have support when doing something for the first time, especially coming to a trauma clinic. 'But I do just need to check with you, Tom,' I said, 'is it okay for Albie to be in with us while we discuss your problems?' Tom nodded without speaking, but reached out his hand to rest on Albie's arm, which helped reassure me that Albie was a valued person, not a threat. I nodded back, deliberately echoing his movement to validate his agreement. 'So, let's start our discussion today with the two of you, and then, Albie, if I might ask you later to step out for a bit, maybe get a cup of tea? It's important to the work that I have time with Tom alone. How does that sound?' The two men exchanged glances, then chorused, 'Okay.'

I started by focusing on the symptoms. 'So, Tom, your GP says you've been having nightmares.' Albie was the first to respond, which I thought was interesting. 'That's right – dreadful, really, thrashing around like mad, crying, shouting at the top of his lungs. The first time it happened, I thought the neighbours would be round any minute, thinking he was being murdered.' Tom gave him a look I couldn't quite read; I was unsure if he was glad or annoyed that Albie had spoken. Perhaps this was just their way of being in company with others. Then Tom picked up where Albie had left off, his voice low. 'It's carried on for a few months, so I went to the doc . . .' He trailed off, as if he had run out of words. Albie leaned forward, his voice dropping a little. 'He was in the war.' I presumed he was referring to the Second World War but waited for Albie to say more. He continued: 'He hasn't told me much . . . I guess it was sod's law he was shipped out to Singapore right before it fell. Landed up in Changi for three years.'

I didn't know much about Changi then, other than having a vague idea that it was an old prison carved out of the tropical swamps in Singapore. The Japanese had converted and expanded it during the war, holding their Allied prisoners in notoriously brutal conditions. I used a soft 'wow' to convey that I understood how dreadful that experience might have been and to signal that I was ready to hear more. 'And have you had these nightmares for years?' I had to check the GP's letter was accurate. 'No!' Tom blurted, with some emphasis. 'Not at all . . . I never had them before. I don't think about that time. I'm fine. It's not . . .' Once again, his words faded, and as before, Albie stepped in to help. 'That's the whole point, Dr Adshead – he sleeps like the dead, you know? This isn't like him. It came out of the blue. It's so upsetting.' He looked anxious, and Tom looked forlorn, and I felt their sadness.

I also thought this might be a good time for Albie to step out for that cup of tea and maybe get one for Tom, too. Before he left, I told both men that I wasn't going to ask about the content of the dreams, not today. Tom seemed not only taken aback, but also relieved, as so many patients do when faced with what I was starting to think of as 'the survivor's dilemma'. They've come to a place where they will have to talk about whatever they most fear, but engaging with distress, especially if you've been trying to avoid it for years, is not easy. I was discovering that many trauma survivors didn't show up for appointments or would drop out of treatment after the initial session.

When Albie left, I checked again with Tom to see if he was happy for him to be with us and if things were okay between them. 'Never been happier,' Tom replied firmly. 'Not something I ever expected,' he added. 'I was in my fifties . . .' I left it there. I didn't need to ask any more about their relationship at this stage; I just wanted to establish whether he felt safe.

'So, your nightmares came out of the blue?' I deliberately echoed Albie's phrase to allow Tom to decide if it was accurate. He nodded. 'And they always wake you?' Tom grimaced, nodding again, then reiterated that he usually 'slept through anything' – rephrasing Albie's 'like the dead', I noticed. I asked about other sleep events, like ordinary dreams. Conscious of his age, I also enquired about needing to get up to use the loo at night and whether that was disturbing his sleep. He said he had no real complaints and had always slept through the night, until about six to eight weeks ago, when all this started.

'And there's nothing you can think of that set the nightmares off? A movie? A meeting with someone from the past?' Tom shook his head, telling me that the first one occurred late one Saturday night. It had been a fine, sunny day, as he recalled. 'I got up late, read the paper, did the crossword and helped Albie with some gardening, planting tomatoes in the raised beds.' Then he took the dog out for his usual walk on the common and watched some of the village cricket match. Albie made dinner for them. 'He's the cook,' Tom explained. 'I'm hopeless. And then it was bedtime . . .' His voice fell away again. I could see he didn't want to say more about the substance of the nightmares yet, and that was fine with me.

To signal this, I decided to ask about something else: 'What kind of dog do you have?' 'A black lab,' he told me. 'He's a love, five years old, but high-spirited like a puppy. Always into mischief – chewing our slippers, chasing Albie's sixteen-year-old cat around whenever she's good enough to make an appearance, the little diva. We dote on him. Monty.' 'Monty?' I asked, not assuming anything. 'Like the general, you know? He's in charge around our house. We just follow orders.' That made me laugh, and Tom joined in.

My old mentor, the psychotherapist Dr Murray Cox, often stressed the importance of a lightness of touch when talking about issues that weigh on people; it can help show them that you are open and ready to hear anything they have to say. I've found that this ability begins with gaining control of my own responses and not reacting to things in a way that might cause the speaker to shut down or run away. Over time, this communication of willingness becomes more nuanced, and humour definitely plays a part. I was relieved Tom and I could laugh together, conscious of needing to build more rapport with a man who came from a generation that might be suspicious of 'head shrinkers'; perhaps this was even truer of someone with a military background, although I didn't want to assume that. What I noticed was how little Tom had revealed of himself up to this point. I'd learned more about his dog.

I glanced at the clock – our hour was half over. I needed to get more backstory at this first assessment, so I asked how often the nightmares occurred. 'I don't know. Weekly, maybe two or three times a week?' I made a note. This frequency was familiar to me, having seen other people with PTSD.

Tom looked uncomfortable. 'I told the doctor I was starting to be afraid to go to sleep. He gave me something to help with that.' He named a benzodiazepine, the usual sedative GPs prescribe for sleep disorders. 'But I don't like the effect. It makes me muzzy in the mornings, and it doesn't really stop me being anxious about the nightmares . . .' Then he offered: 'They're always about Changi. I don't know why they've come crashing in like this now – that was another life.' I said nothing. He shifted in his chair, glancing at the clock on the wall, and sighed. I had the impression he was frustrated, as mystified as I was about why such old memories were surfacing. His use of the word 'crash' had me imagining his trauma appearing in his dreams like a Japanese kamikaze pilot hurtling down from the heavens, attacking him 'out of the blue'.

I decided to offer a question he would be able to answer with confidence: 'Can you tell me something about your ordinary life, before the nightmares began?' 'How do you mean?' 'Just anything you'd like to tell me, to give me a picture.' I left it as broad as possible; it's always interesting to hear where people choose to start their story. 'Um, I'm retired . . . Albie and I live together in Haslemere. It was my parents' house. I moved in there with Mum after my father passed away, back in '64 . . .' He lapsed into silence. 'How old was he?' I prompted. He met my eye. 'Oh, about the age I am now, I guess.' I sensed he'd not considered this before, and waited to see if he wanted to say more. Instead, he shrugged, dismissive. 'We were never close. Whereas Mum . . .' He looked away, out of the window. Even if time was short, I didn't want to miss this opportunity. 'Mum?' He smiled then, a mixture of love and sorrow evident in his face and tone. 'She was the best. I miss her every day. Fabulous cook. Looked after us – and everyone else around, too. The biggest heart . . .' It was a bare pencil sketch of a loving carer, but at this point, it was enough for me to understand a little about who the young man was when he was sent to Changi.

Tom changed gear, telling me that after coming out of the army, he trained as an engineer and got a job with an electronics firm. 'Worked in the one place all those years, until I retired a few years ago. I thoroughly enjoyed it. A good job. Good people.' 'And outside of work?' I prompted, hoping for a fuller picture. 'Any hobbies?' 'Oh, I used to have loads when I was younger, outdoor things – sport . . . you know . . .' 'Sport?' He shrugged. 'Rock climbing, caving, snowboarding sometimes . . . that sort of thing . . .' 'Sounds a little risky,' I said, with a note of gentle interrogation to show I wasn't judging him in any way. 'I suppose I was a bit of a daredevil in my younger days,' Tom said, his self-deprecating tone belying the very idea. 'A daredevil?' I repeated, intrigued by the word, but he didn't elaborate.

Instead, he went on a tangent about his enthusiasm for DIY. 'We've put in new bathrooms, kitchen cabinets . . . The old ones were so outdated, you see. Mum and Dad never did much to the place. Just installed a new fireplace mantel, too, the other week. Beautiful – restored walnut, sort of art deco, with lovely chrome details . . . I need to do the roof next, but Albie says best not . . . We'll get someone in . . .' That was interesting – he was opening up a little, perhaps revealing some vulnerability. I wanted him to keep talking.

'Anything else?'

'Well, there's my bike. We ride every weekend.' My first image was of men on ordinary bicycles, cycling side by side down leafy lanes, but then he explained: '1960 Triumph TR6. She's my pride and joy. Always promised myself I'd get one, and finally, about fifteen years ago – well, that's how I met Albie. Same club. Turns out he knew a man who knew a man who was getting rid of her at a reasonable price . . .' He smiled for the first time since we'd met. 'And we've been together ever since.' The broad sweep of his hand in the air encompassed the motorcycle, Albie, Monty the black lab and the ageing diva cat, all happily ensconced under one unrepaired roof.

'You've never seen a psychiatrist before? Even after you were demobbed?' He shrugged. 'People did. But I was fine. I mean, it was a bad time, all that, but I was healthy, things were getting back to normal. You don't want to get into it or, you know, put those things on people. You have to get on with it.' It reminded me of the Jorge Luis Borges story about a man who is cursed with remembering everything – that powerful idea that in order to live, it is necessary to forget.[2] I also thought of several people I'd seen in this room who had shared with me how they worried their trauma experiences might be contaminating and burdensome to others.

I had one last question, conscious Albie would rejoin us soon: 'Tom, can I ask if the nightmares have affected your relationship with Albie?' I needed to know. Relationship breakdown is very common in traumatised people, and sometimes the consequent loss of support leaves people isolated, which then makes everything much worse. Tom paused, before answering: 'I think . . . this is hard . . . frightening. He's sleeping in the other room . . .' I wished I knew what he wasn't saying when his voice ebbed away like that. His consistent inability to complete his sentences suggested something was blocking his capacity to tell his story. I didn't know whether the GP had done tests for dementia, and made a note to ask.

I stepped outside to call Albie back in, asking him quietly as we returned if there was anything he wanted to tell me, something

Tom might not have been able to say. We could have a phone call, if he wished. He said he'd let me know, oddly short of words when he wasn't in the same space as his partner. Before they left, I told Tom that although his story was somewhat unusual, there were things we could do. I invited him to come back for a few more sessions. When he agreed, I asked if he would consider doing some homework before we met again.

His face fell. 'Homework?' I explained that I'd like him to start keeping a diary of when the nightmares happened, along with some brief descriptive details, whatever he could remember of them when he woke.

'A diary?' Tom sounded sceptical, as some patients do in the face of this suggestion. Albie inserted that 'Tom's more a data person, an engineer.' Used to their routine now, I said nothing, knowing that Tom might jump in to add something, which he did. 'That's right. I'm not much of a writer . . .' A lot of people say this when given the task of journalling in exposure-based therapies, but if they don't do it, success can be limited. Sometimes I have to find ways of framing it so that it seems less alien or difficult. 'Could you think of it as data collection, perhaps? Like for a project that you're building?'

He brightened up immediately. 'Oh yes. I can do that. I do it when planning anything. Remember all the research we did for the bathroom?' He turned to Albie, who rolled his eyes, and we laughed again. Another light moment, and a hopeful one, too, because it showed me that Tom was ready to take a chance on psychological work. Already he might have felt he could have a little more agency over these mysterious nightmares which were so upsetting for both of them.

It was time to go. 'Do you have any questions for me?' I enquired. Tom hesitated, glancing at Albie, then asked: 'How long will it take? I mean, to stop the nightmares?' I was honest, telling him we'd have to take it week by week.

———

Before our next session, I read about the Allied POW experience in Singapore, in preparation for whatever Tom might bring back. I thought filling in some context that he might not include could be helpful.

I soon learned that Changi wasn't a prison as I understood the term, but rather a fifteen-mile-wide area of land, a sprawling amalgam of seven different camps clustered around a former civilian prison at the eastern end of Singapore island. After the British defeat in February 1942, nearly 100,000 Allied soldiers – and many civilians as well – were taken prisoner overnight. The higher-ranking officers were soon dispatched to camps in Japan, while the rest were set to building barbed-wire fences, forced to be the architects of their own incarceration. Changi was remote from the city and stood on the edge of a forest; the name derived from a tribal word for 'tall trees'. It was built on a former mangrove swamp, and people described it as a marshy tropical hellhole, swarming with insects, rodents and other disease-causing pests.

However, at least for the first few years, life in Changi sounded somewhat bearable, when compared to the experiences of the men sent to camps in Japan, where conditions were even worse. Tom and his fellow prisoners had food rations – a cup of rice daily, some fish maybe – and they were allowed to plant vegetables for their own consumption. Prisoners could even send a rare letter or postcard to their loved ones. Their captors strictly limited these to twenty-four words, a small deprivation, but one that struck me as harsh. Another kind of silencing. What can you tell someone in two sentences, other than perhaps offering bland reassurance and some abbreviated expression of affection?

It was always a physically gruelling existence. In extreme temperatures, prisoners were assigned to large-scale projects, such as building a new airport runway nearby. Other prisoners were sent further afield, including to help construct the notorious Burma–Thailand Railway, known as the Death Railway due to the treacherous working conditions and immortalised in David Lean's film *The Bridge on the River Kwai*.

But there were some remarkable signs of ingenuity and resilience at Changi. I read about a makeshift lending library and even a camp 'university'. Boxing and cricket matches were not just tolerated but encouraged, presumably because keeping fitness up meant more productive workers. Some of the soldiers had been able to bring musical instruments to Changi when they were captured, and there were concerts and singalongs, at least in the early days. I found one especially moving and memorable image of a battered old violin (now held in a Melbourne archive), etched with the words, 'We'll never get off the island.'[3]

Later, as the war started to go badly for their captors, any 'privileges' were withdrawn, including all pens and paper. Food supplies were reduced, and life in Changi deteriorated. The Japanese did not observe the Geneva Convention and corporal punishment was commonplace, especially if anyone flagged while working on building projects. I didn't linger over the details of the diverse and inventive kinds of torture that were used, choosing not to think about them. Given my other work with perpetrators of violence, this avoidance might seem strange, but there was something about the helplessness of these thousands of men that made me want to detach. It was also disturbing to me to learn about the prevailing cultural belief in Japan's military at the time that anyone who surrendered was beneath contempt, meaning that their prisoners were dehumanised. I imagined that this was what allowed those who ran the camps to justify such extremes of cruelty, an idea sharply reminiscent of some of my forensic patients, who believed that their victims were 'things' to be denigrated or extinguished.

As the camp's population swelled with captives from all around the Asia-Pacific, hunger became the biggest issue. By the end of the war, a third of the Allied prisoners had died. In reference books I found sepia images of skeletal figures, dressed in loincloths and with dark, sunken eyes and jutting ribs, packed six or eight into cells meant for one. It was hard for me not to think of the Nazi concentration camps, despite the stark distinction between serving soldiers held as POWs and a civilian population incarcerated only because they were hated. One of the buildings in the encampment, the Roberts Barracks, was designated as a hospital, but it was ill-equipped to deal with the volume and nature of the needs that arose. I imagined that those who survived Changi were lucky enough not to get sick or by some miracle got admitted to hospital long enough to carry on living, as in the story Primo Levi tells in his epilogue to *If This Is a Man*,[4] while those outside perished in their thousands.

For our second session, Tom came independently, which was pleasing, as it suggested that he felt comfortable with me. He told me Albie had dropped him off and gone to do some shopping, and would collect him afterwards. I was glad he was in such a stable, supportive relationship – this was not always the case for the patients I saw at the clinic.

We sat down facing each other in the small space, a few feet apart. Right away, Tom showed me a small, ruled notepad of the type favoured by reporters back in the day. 'Homework,' he said, with a diffident smile. He flipped open the cover, clearing his throat, and I watched as his eyes darted across the scrawled script. 'Set this by the bed, as you told me,' he said, 'but then the pen rolled away, and I was scrambling around in the dark, and I . . . and then . . . well, things got muddled, and I wanted to get it right, so I had to wait a few more nights, until . . .' I wasn't in a rush to have him read me the details he'd managed to record, nor was I there to analyse or interpret anything. My role was more like a coach, encouraging him to say what he could and interjecting the odd 'That makes sense' or 'Well done' as he went on. The significant thing was that he'd been open to looking at what he feared.

As that thought occurred to me, Tom closed the notebook, setting it on his thigh and covering it with both hands, pressing down a little, almost as if to smother the contents. His eyes avoided mine, fixed on something or nothing just beyond my left ear. 'I don't know. Maybe there's . . . no point. None of it . . . it didn't really . . . add up. Make sense or . . .' His high cheekbones were a little flushed, and I saw his hands were shaking. I would need to help him manage the anxiety he was experiencing, without appearing to be running away from the task at hand. I said, 'Tom, could you perhaps start by telling me about just one element, some detail that felt particularly upsetting or surprising?' He stared at me, saying nothing. I cast my mind back to the last session, wondering if I could have missed the tremor in his hands before and whether that might indicate something other than anxiety. Then Tom spoke, his voice halting and flat. 'It was so . . . I woke up screaming again, covered in sweat, feeling sick to my stomach. Albie came running in, and I thought I was . . . you know . . . It felt like it was still happening for a few minutes. Took me a while to settle. I don't think I slept after that.'

'Can you tell me a little about what you wrote down?'

He opened the notepad and started to read and interpret his notes for me. 'I'm in Changi . . . I think I'm with Ben, near the edge of the forest, we're stacking logs . . .' He looked up from his notes, eager to supply some context. 'Ben was our captain. He was the most senior man left from our unit . . . Such a kind man. Salt of the earth. We were in the same cell for a while, but we all got moved around a lot, chopping and changing . . . I'd been there ages by the time he came in – he'd been on the Burma railway?' I indicated that I understood the reference and felt grateful for my brief research. 'Probably just had ten years on me, but he was like an old man. They'd tortured him out there, people said . . . He had a limp, couldn't move his right arm, hair going prematurely grey . . . But he'd do anything for you. Shared his rations more than once . . . always telling me to keep my chin up.'

Tom continued to talk about Ben's character, more voluble than he'd been so far. 'He would always tell us we had so much to look forward to, our whole lives in front of us, that sort of thing. Hopeful words. How could he have any hope left? He was just one of those people . . .' I stayed silent, not wishing to intrude on this poignant memory. He shook his head. 'You should have seen his smile. It was like a child's . . . full of joy, warmed you to see it. We called him Big Ben. 'Cause he was the oldest, you know? We all got nicknames . . .'

'Did you have one?' The logical question. 'Um . . . various . . . Some of them called me Hercules for a while, as a joke. I mean, not a joke exactly. I was one of the youngest, maybe fitter than most. I don't know . . . I was up for anything . . . Give me the toughest task, I'd do it.' The daredevil. I was intrigued, given that

Hercules was fighting to recover his immortality. But for the time being, I let these things lie. I wanted Tom to say more about the contents of his nightmare, if he could.

'Where was I?'

'With Big Ben?' I prompted. He took a deep breath. 'I have to get him to Roberts.' He didn't pause to explain that reference, but from my reading, I knew he meant the camp hospital. 'The thing is, he's a big man – almost a foot taller than me and twice as wide.' I wasn't sure if he realised that the picture he was painting of this greying, tall, broad-shouldered man might also be a description of Albie. But he seemed oblivious to that, caught up in his account. 'But he'd lost so much weight, it was dead easy for me to . . .' He glanced down at his notes. 'I see blood dribbling out of his ear . . . I think they hit him round the head when he fell, that was the usual thing . . . So I lift him in my arms . . . like a sack of feathers, he's so light . . .'

His voice had started shaking, but I didn't interrupt or try to soothe him; it was important that he felt he could manage this and didn't need my help. For a long minute, he didn't say any more, but then he cleared his throat and consulted his notepad, tracing the words with the tip of a long forefinger: 'Eyes rolling back in his head . . . lips blistered and swollen, tongue lolling out . . . It's horrible, he's dying . . . then he vomits. Black bile. All over me. Black bile and blood.' He closed his eyes, his voice dropping to a whisper. 'A lot of people vomit before they die.' When he opened his eyes again, I saw they were brimming. I didn't ask if he wanted to take a break. Tearfulness when you feel grief or fear is healthy and normal, and I very much hoped he would keep going.

But that was enough for the day. 'Can we stop?' He took out a handkerchief and wiped his face with a rough scrubbing move-ment, almost, I fancied, as if he wanted to erase whatever was inside his head. 'Of course,' I said. 'I know this can be uncomfortable, Tom, but a nightmare can be a sign that your mind is working on feelings that need thinking about. Painful memories aren't an illness that needs to be cured. If they come up in your sleep, it may mean new thoughts and appraisals are happening in your mind. Does that make sense?'

He frowned. 'I don't know why it's all coming up now . . . I'm annoyed, to tell the truth . . . After all these years. I put all that behind me. Got on with things, with life . . . I've never . . .' His incapacity to finish a sentence was getting more pronounced. What was it that Tom had never done, or thought, or said? He waved at the page of writing in his lap. 'What does all this do? I mean, will it make the nightmares stop?' I said I had known it to bring relief to people, and if he were willing to keep going, it might do the same for him. Organising memories of painful experiences into stories can help people feel more in control, I explained, because they get to choose the words, which can make them feel less overwhelmed by their emotions.

I was also thinking about studies involving people with trauma who were asked to write about the events that distressed them, which is known as 'expressive writing'. Researchers of this kind of treatment report that people who do this have reduced stress levels, enhanced well-being and improved immune function. The mechanism is not clear, but it may be that putting memories into words restores a linguistic sense of agency, and communicating those emotions to others reduces their intensity, thereby 'detoxifying' the psychological pain.[5] It's also possible that writing about painful emotions can change the way they are organised as memories of threat. A recent study by Dr Charan Ranganath backs this up,[6] demonstrating how the act of remembering is an act of imagination that will alter a memory over and over. He likens memory to a photocopy of something – a flyer, say, that is copied repeatedly. It will distort over time, until the copy of the copy of the copy becomes something almost unrecognisable.

I had another question for Tom: how many times had his nightmare recurred since our last session? Tom checked his notebook: just twice. I noted that this was a little less frequent, and Tom relaxed a bit. 'That's good then, isn't it?' And we made another appointment for the following week.

———

At the time I was meeting with him, there weren't a lot of therapeutic interventions for nightmares, possibly because the people who had them tended not to seek help, and because there was some professional debate about how best to treat them. Nightmares do not always feature actual events (although they can do), but they often act as a vehicle for emotions like fear, helplessness and terror. They have been recognised as a feature of post-war stress since Homer's time. Shakespeare provides a vivid account of them in *Henry IV, Part I* that is pretty clinically accurate to this day, as a returned soldier's wife watches her husband sleep and laments how his 'spirit . . . hath been so at war' as he slumbers.[7] However, the cause and function of such disturbances while we sleep are still obscure.

There was a new approach to post-traumatic nightmares just being developed back then. Known as Imagery Rehearsal Therapy, or IRT,[8] it showed promise in reducing nightmare frequency and distress, even after only one or two sessions. I didn't know that much about it, but I was aware that the aim was to help people revisualise their nightmares. They were asked to keep a diary (like Tom's) and then alter aspects of the narrative and imagery by writing out a new scenario to make them more positive: the lost child is found playing in his room; the plane lands safely. They are then asked to read through the 're-scripted' dream story before bed each night until the new images take hold. If I could help Tom feel a bit less scared of the nightmares by using the diary, and if we could come to some understanding of why they were invading his sleep fifty years after the war, then I might refer him to a colleague who was testing IRT in their lab, and perhaps Tom could become part of a trial.

———

I discussed this plan with Tom when I next saw him, and we continued to meet weekly over the next month or so. We would return to his time in Changi and to the fragments of his nightmares that he was able to record in his diary. They decreased in frequency, but the images in the nightmares remained very 'live', and he would still wake up screaming and shaken.

Gradually, we pieced together a rough narrative. Although many of us may think of nightmares as a series of terrifying and threatening images, a chaotic, flickering horror film in our mind's eye, our recall of them will often involve the other senses. Tom had already told me how it felt to carry his friend's featherweight body in his arms, and he went on to tell me how hot it was, describing how his clothing was sticking to his body as he tried to run with his burden and the acrid smell of the man's breath. There were aural sensations that came up for him, too: the shrilling native birds and buzzing insects in the forest, and the harsh shouts of the Japanese soldiers behind him, using threatening words he couldn't understand. I sometimes felt like I was with him, it was so vivid.

As he got used to the diary work, he became more able to articulate the emotions attached to each sensory impression and describe his feelings on waking. There was a confusing alternation between hope that if he could make it to the hospital in time, they'd be all right, he told me, and a terror of being shot before he could get Ben safely there. But sometimes Tom was too shaky to say much, and he'd set his notes aside, changing the subject to talk about what was going on for him day to day, and I was happy to listen. He might discuss how he and Albie were planning a bike trip around Wales in the new year, or how he was busy building sets for the am-dram society in the village, who were readying their annual Christmas pantomime; Albie had been cast as a genie and was obsessed with putting together an elaborate costume. Only once did Tom arrive late, full of apologies, telling me he'd been raking leaves all morning and had lost track of time. I was touched when he presented me with a perfect oak leaf of a spectacular orange hue. If we had been seeing each other long term, I might have said more about this gift, inviting him to help me understand its meaning and why he was giving it to me now. But in the time available, I couldn't get into everything and simply offered thanks for his thoughtful gesture.

———

One day, several weeks in, there was a shift. Tom was reading from his notebook, and I glanced down at his usual illegible scrawl, a series of words and short phrases in a column, always more like a laundry list than a journal. I could read one word upside down, the last on the list. It was in capital letters: 'HANDS'. In his most recent version of the nightmare, he was climbing a hill, then tripped and dropped Ben, who rolled down the slope. I could see what horror this prompted in him. 'I run down, roll him over,' Tom half whispered. 'His eyes stare up at me, but they don't see. He's gone. I've killed him. He's dead. And then I realise – I can't get up.' His breathing was getting short, his face reddening. 'They're right behind us. I hear them shouting, getting closer . . . and Ben is gone. It's just me, lying there – and if I don't get up, I'm dead.' There was a silence, which I think we experienced differently: his I'd come to recognise as the dumb silence of fear, while mine was the tactical silence we learn as therapists, a non-judgemental wait for whatever might need to emerge. We sat and breathed in unison for a bit, Tom following my inhale and exhale as if we were practised at it. 'And then?' His face was blank. 'Oh. Then I woke up.' I didn't press him about what I'd seen in his notes, about the significance of the word 'HANDS'. Whose? Ben's? The enemy soldiers'? I would leave it for now, but I thought it would be important to come back to this in future.

On a chilly day in early November, Tom arrived at the clinic, quite obviously unhappy. He scowled as he took off his scarf and hat, and I noticed he had a paper poppy pinned to the lapel of his wool coat – the symbol of remembrance for those who died in war that blossomed in buttonholes across Britain at this time of year. It fell to the floor as he hung the coat on the hook on the wall. I scooped it up and handed it back to him, and he played with it, turning it over in his hands while we talked. Was he involved in London's Remembrance Day activities? Did he usually attend the service for veterans in Whitehall? I was curious as to whether his mood was somehow connected to this.

'No, not really . . . I don't go in for all the pomp and ceremony, the wreath-laying and the royals . . . But I always raise a glass to . . .' he stopped, his voice catching. Absent friends. I finished his thought in my mind but resisted saying it aloud. Raise a glass to our absent friends.

'This time of year, I always feel . . . I don't know. Heavy? No, sad. Things dying. Bare trees. And so many dead, so young. The

waste. Not just my friends, although I do think about them . . .
Good men, people like . . .'

'Ben?' I guessed, then chided myself internally for doing so.
By this time, I knew better than to fill in a patient's linguistic gap,
but his way of speaking was so tentative, I couldn't help it. I was
finding myself starting to empathise with Albie. It was difficult not
to prompt someone who trailed ellipses behind his every thought
like a blank sky banner behind a light aircraft.

He looped back to my original question. 'We've gone along to
the Cenotaph once or twice. But in the end . . .' With a frustrated
gesture, he burst out: 'I should be happy, Doctor. I'm trying.
Everything's going well with us. My life is good. The nightmares
are getting better . . .' He flicked his notepad. 'Only three this
month. But I feel . . . on edge all the time. And then . . . the cere-
monies, all the fanfare – it doesn't help.' I was looking at what he'd
done to the poppy in his hand, folding its petals inward, rolling it
around in his fingers until it was reduced to a paper berry; it was
a mutilated, crushed fragment of remembrance. He cleared his
throat again and sniffed, trying to collect himself. The edge of
tears was as far as he would let himself go.

'Tom,' I said after a bit, 'what if we leave your notes to one side
today and look at something else?' He eyed me, uncertain where
this was going. I shared with him my continuing preoccupation
with why the nightmares had begun. I suggested we could use
the session to revisit that summer Saturday, which he associated
with their start. Could he take me back to that ordinary day he
described in our first session, when he did a little gardening and
watched the cricket? 'Would you be able to walk me through it?
Everything you can remember.' Tom was a little bemused by this
request, but we had built up a good enough rapport by then –
good enough for him to share his tearfulness and anxiety with me,

at least. I was hopeful that perhaps he might recall more this time, with his memory like the aperture of a lens, widening in the light of our connection, letting in more detail and emotion. I might even get to ask about the meaning of the note 'HANDS'.

Starting in the morning, on awakening from a sound sleep, he'd read the papers, he told me, then did the *Times* crossword, a daily habit. Nothing at all disturbing or unusual there. Albie made breakfast: the full English, a Saturday treat they both loved, with baked beans and mushrooms, the lot. Then into the garden, doing the raised beds . . . oh, and he'd cleaned out the stone birdbath, too. After that? 'As I said before, I walked Monty while Albie was doing the weekly shop, then came back home. That's all.' 'Tell me about the walk,' I asked. 'I don't really remem— Well, we must have done our usual route, out along the high street and then cutting through the pub's car park to get to the common.' He thought for a moment, then recalled that there was a footie match on TV that afternoon – he had heard the telly and the cheers and groans of the fans in the pub. 'Did you stop in, maybe watch a bit of the match?' I asked. He waved his hand. 'Not my sport these days . . . It's all become a bit too tribal, you know? It wasn't always like that.'

As if to prove his point, he was able to remember seeing a couple of lads round the back of the pub, a few pints the worse for wear, wearing different-coloured team shirts. One stabbed the other in the chest with his finger, shouting something. Tom couldn't hear the words, but it looked like they were getting right into it. When a few other fans rolled out to join the fray, jostling and jeering, he picked up his pace, urging Monty along towards the peace of the common. So far, so ordinary. I continued on his walk with him.

Emerging through the trees onto the edge of the common, he was glad to see the cricket match was still on, and he stood and watched for a bit, while Monty nosed around in the grass, on the scent of squirrels or rabbits. Suddenly, the lead jerked hard out of his hand, and the dog was off, chasing down something he couldn't see. Tom loped after him, calling his name in vain, ducking under tree branches, his eyes on the end of the slithering lead, just beyond reach in the dark grass ahead. Then he tripped over a bulging tree root and took a little tumble – not flat on his face or anything, though. 'No drama,' he assured me. He landed on his knees, breaking his fall with his outstretched hands. He tried to get up but found himself at an awkward angle, on a bit of a slope, with tree roots hemming him in – and as Monty came snuffling back to him, he found he was unable to stand. He was stuck.

The cricketers were changing ends for tea; he could hear their voices a little way off. 'Hello?' he called out, embarrassed, and soon heard a shout in response. A young man he vaguely recognised, sunburned of face and red of hair, ran across, all kindness, pulling him to his feet and enquiring if he was all right. 'This all happened in seconds, Doctor. Honestly, I didn't even mention it to Albie. It was nothing . . . I just . . . I just saw . . .' At this, he looked crestfallen and quite vulnerable. He dropped his head.

'You saw . . . ?'

'My hands,' he blurted.

'Your hands?' I was so glad I hadn't needed to ask. Now we were here, at his note from the last session. 'Look at them,' he mumbled. Head bent, not meeting my eye, he spread them out, palms down, fingers splayed across his knees. Strong-looking hands, I thought, the skin tanned and tough, with prominent bluish veins. I wasn't sure what to say. 'Have you hurt yourself?' I gestured to a thumb, noticing a smudge of black under the nail, likely due to an errant hammer blow during one of his many DIY projects.

I sensed Tom's irritation that I wasn't understanding him. 'See those spots?' His voice was full of distaste. There was a scattering of pigmented liver spots across the backs of his hands – the kind we associate with age. Tom sighed. 'That boy was so young, so healthy, so caring. He knelt, his hand on mine, asking if I was okay . . . then very gently lifted me up and helped me stand. And suddenly, I saw the difference – I saw how he must see me. I'd never thought of myself like that before . . . an old man in the dirt . . . helpless.' The language was so like his description of the dream, in which he was in the lad's place, a heroic young demigod hoisting Ben into his arms to try and save his life. I had a thought about the fight at the pub he'd passed on his walk and wondered whether the aggressive shouts of the football fans, indistinct but threatening in their way, had recalled those of his Japanese captors, which he'd described to me before.

I was curious as to why he'd not mentioned that fall to Albie, when it had affected him so much. 'He'd only worry . . . and nothing was broken . . . I was fine . . . I put it behind me.' Again, it was the same language he'd once used in reference to his experiences in Changi. I recalled how some psychologists see the word 'fine' as a telling acronym for 'feelings inside not expressed'. It was becoming inexorably clear that something had broken in Tom that day. But I still didn't quite understand the significance of 'HANDS' in his nightmare.

'Tom, do you remember me telling you how nightmares and dreams are a way for our brain to process memories and fears? What does that mean for you here?' He took out his notepad, urgently flipping back through the pages until he came to the list I'd seen before, his forefinger hovering over the word 'HANDS'. 'So, I drop Ben, and they're coming to get me and . . .' He broke off, shaking his head from side to side, refusing to go further.

After some seconds, I nudged him to try and finish the sentence. I thought he could do it. I let him gather himself, before asking quietly, 'What's going on for you now, Tom?'

'I thought . . . I'm thinking, I'm afraid . . . we won't make it. I look down, and my hands . . .' He made a little sound, a strangled gasp. 'They're . . . they aren't like they were back then . . . They're like this.' He held up his hands again for me to see, his voice faltering. 'Look at them. Old hands. I know I will never survive this place now I'm old . . . I'm seventy-two. I can't even go up and fix the roof any more, can I? So how will I survive the camp if I'm an old man?' We sat for a moment, contemplating this together, before I observed: 'In your nightmares, Tom, when you tell me about them, you often speak in the present tense . . . but, of course, the past is gone, it's history. It doesn't exist, except as the memories you recall.' No response; he was still gazing down at his hands, turning them over like new-found objects. 'And it strikes me, Tom, that the phrase has another meaning.'

'What phrase?'

'An "old hand". Isn't that a way to describe someone with experience, even expertise? Someone who perhaps knows how to climb a mountain, race a motorcycle, mend an engine or install a bathroom, coax plants to grow or record a life full of memories?' It's always valuable for a therapist to notice any play on words, cliché or new metaphor, not least because these linguistic devices can reveal an important fresh thought. It also shows that you are paying attention, and my 'play' with his words seemed to shift something in Tom. He allowed himself to smile at me, taking pleasure in the leap I was making. 'An old hand. Well, that's another way of looking at it, Doctor.' He put a firm full stop at the end of the sentence.

———

After that session, things changed for the better. The nightmares were reduced to the point of disappearance. Tom told me he had a greater sense of control and felt less terrorised by them when they did come. We met a few more times, and he was able to talk more about the sadness that came when he thought of the lost years of his youth. He recognised that he had been battling for some time against the natural physical changes in his ageing body. We agreed that an increased sense of vulnerability may have dislodged 'data' related to his time in Changi into his dreams. The former soldier, no longer an invincible youth or risk-taker, was not conquering any more.

Around this time, he reported a rare argument with Albie. They'd bought a new piece of art together, and he was about to hang it in the sitting room, when Albie insisted he shouldn't climb a ladder. Tom had exploded – what a ridiculous thing to say. 'I'm a man who's climbed sheer cliff faces for a hobby, you know.' His face reddened as he recalled the argument. But Albie prevailed, and Tom grudgingly admitted to me that his balance wasn't what it used to be. 'I still felt humiliated, though.' I murmured something like 'I understand', and he briefly lost his temper with me, cutting me off with a blunt comment: what was I, thirty or something? How could I possibly understand? I agreed that ageing was not something I had experienced yet, but I knew it was natural and inevitable. I thought, but did not say, that perhaps he resented my youth, and even Albie's, although he hadn't said as much. I also felt he wanted to get me to appreciate his age. Only then did it occur to me that maybe the perfect autumn leaf he'd given me had signified this. Or maybe it hadn't. One of the biggest legacies of psychoanalysis is that we've come to think all actions and images must have a deeper meaning. But sometimes a leaf is just a leaf.

Our one brusque moment passed quickly, but it allowed us to begin a conversation about the transitions at different stages of life, and the shifts in identity that occur at each one. After all, we both knew the experience of being children, of becoming young adults, and then older adults with jobs we enjoyed, and this allowed us to find common ground when thinking about how people develop through their lifespan. I shared with him some of the psychoanalyst Erik Erikson's research into what he termed the eight stages, or 'building blocks', that contribute to our psychosocial development.[9]

Erikson proposes that although older age may be associated with some despair around the loss of autonomy, or a sense of things not done or said, it is also a time of integration of our experiences, of reflection and possibly greater appreciation of our mastery of earlier challenges, if they have been part of our journey. They certainly had been for Tom. It seemed to me that he was reluctantly beginning to accept that we all become more dependent on others as we age but are free to consider how we might still enjoy our liveliness. The sadness I noticed during our first meeting hadn't been erased, but perhaps transformed into moments of melancholy, which Italo Calvino so beautifully defines as 'sadness that has taken on a lightness'.[10]

Despite the difference in our ages, it seemed Tom and I both held a limiting belief that needed challenging. At thirty, I was busily forging my identity, professional and otherwise, and back then, I'm pretty sure I thought that at a certain point I'd be 'done' and that older people were 'set in their ways'. Tom's story showed me that this was not only a false but a dangerous presumption. We are always changing our minds, right through to the end of our lives.

This is one of the most essential precepts of trauma therapy, and it can be a source of both tension and hope.

I think Tom had come to our first meeting also feeling wary about opening the door to the totality of his experiences in Changi, but there had been no need to get him to recount or relive all the torment he had known there. Much as physical therapy focuses on the symptoms and the specific body part that is hurting or malfunctioning, so, too, in trauma therapy survivors do not have to 'go over everything' in order to transform their pain. Nor did Tom need to move on to additional therapy, such as IRT; he'd been able to change the ending of his story without it, by gaining a better understanding of the source of his fears about ageing. He had recognised that it takes bravery to be vulnerable, having survived an experience where weakness meant certain death, and this would hopefully reassure him that dependence does not equal weakness. Quite the opposite, when we considered his secure, loving relationship with Albie. Love is always a choice to be vulnerable; that is the basis for honest and mature intimacy. He had a solid foundation to build upon.

I don't think I did all that much in this case, other than provide a safe space for Tom to be heard, allowing him to organise some fractured thoughts that he'd been carrying in his mind for decades, which had become quite a heavy burden. Not only did his sleep and frequency of nightmares improve, but I also noticed a change in his speech patterns, with fewer ellipses and a greater sense of agency over his language choices, which allowed him to complete his thoughts aloud.

As we wrapped up our final session, I recalled Tom's early definition of himself as a 'daredevil' and reflected that he had taught me that there was a flip side to my conception of the term. I had previously defined it as someone having an unhealthy appetite for risk – understandably, given my forensic training had been all about helping people reduce their risk to themselves and others. But Tom demonstrated to me that those who dare to venture into the wilds of their mind and heart (at any age) may emerge stronger and more ready to face the next challenge.

It was moving to say goodbye to him. He took my hand in both of his and thanked me with great courtesy and warmth. His story never left me, partly because of his personal courage, but also because he showed me how particular recovery from trauma can be, and that we do people a disservice if we suggest it ought to conform to a set timeline. Trauma is not an event, it is a process, and we should therefore be cautious in generalising about how long it takes to recover. The word 'recovery' implies getting back a lost self, but I was beginning to see, along with Tom, that *discovery* of a new self was more likely, and probably healthier as a goal.

I watched him go, seeing him take Albie's arm as they stepped out into the busy London street, and I considered what might have happened if he'd never had that fall and the nightmares had never surfaced. If Tom had never reckoned with his fears, would his relationship have eventually been strained to breaking point? Or might he have turned to the harmful use of alcohol or drugs, as many people do to anaesthetise their distress? More likely perhaps, he might have developed some physical condition whereby his distressing time as a POW manifested in his body, in the form of chronic pain or digestive issues, for example. However, who can say whether it is inevitable that trauma will find an outlet every time, like a river that must ultimately flow down to the sea? It would be easy for a trauma therapist to believe that, but I'm keenly aware that we know only about the people who come seeking help, and maybe we don't think enough about those who have difficult experiences that they contain and grapple with alone.

The mysteries of the mind are endless and do not tend to resolve neatly like the final pages of a good detective novel. But in this case, it had been rewarding to work out that Tom's vivid nightmares were not simply about his identity as a POW in his youth. Memories of that time had jump-started his mind to process the ordinary fears of a man in his seventies who didn't want to get old and die, who was holding on to his identity as Hercules, battling to join the immortals.

The Refugee

'Got a minute, Gwen?' I was on my feet, buttoning my winter coat, hurrying to leave the clinic for my commute home, when Owen put his head around my door. He was a child psychotherapist, and I always thought the children he worked with must take comfort immediately on meeting him, with his tufty halo of grey-blond hair and warm brown eyes framed by well-earned laugh lines. 'What's up?' He folded his angular body into the chair usually reserved for my patients. 'It's one of our parents – the mum of a three-year-old boy I've been seeing for a few weeks. They're from Sarajevo. Came to England when he was a baby, towards the end of the war.'

A good proportion of the patients coming to the Middlesex trauma clinic, where I was now working, had escaped conflict and persecution. Many were refugees from the former Yugoslavia, a federation that had fractured along ethnic and religious lines during the 1980s. After the fall of communism in 1992, when the European Community and the US recognised one part of the former federation, Bosnia, as an independent state, civil war had erupted. Serbian leaders rejected the move and tried to impose unity through force. As neighbours turned on neighbours, the capital city of Sarajevo was besieged, and Bosnian men were imprisoned in camps. A targeted programme of ethnic cleansing of the Bosniak (Bosnian Muslim) population began.

NATO airstrikes eventually brought the war to a conclusion at the end of 1995, but by that time, the death toll was assessed at over 100,000 people, with more than two million displaced from their homes, many of whom sought refuge in other parts of

Europe. Disturbing accounts of war crimes filled the newspapers, including genocidal attacks such as the wholesale massacre of thousands of Bosniak men in Srebrenica in the summer of 1995.

I'd been aware of all this at some level, but truthfully, I had been preoccupied with my own small world, as I was on maternity leave around that time. By 1997, I was back at work, and had reduced my hours as a consultant forensic psychiatrist at Broadmoor and applied for a part-time post at the Middlesex. I felt I had a lot to learn from the team there. It was one of the few trauma centres anywhere to provide a psychological therapy service to traumatised children and their families. I saw this as an excellent opportunity, because I was now also training to become a psychotherapist. Although all psychiatrists should be able to offer psychological therapy, I now wanted to go further and qualify as a therapist, and I had become immersed in studying childhood attachment and its effects on the developing mind. My new role at the clinic would involve the treatment of adults with PTSD, but I would also do occasional assessments of, and consultation with, the parents of children who had been referred to our team.

Besim, the little boy Owen was seeing, had been referred by his GP because his mother was concerned that he was not talking. 'He's been checked out for deafness and developmental stuff,' Owen told me. 'Nothing wrong there.' I recalled a few cases of speechless children I'd seen while spending time in a child psychiatry clinic during my training. There's a big range of 'normal' when it comes to walking and talking in the early years, but by the time a child is three, you'd expect them to be using a few words. Speech delay is quite common, and there can be many reasons for it, but if physiological causes have been ruled out, one of the first questions is whether there is any stress going on in the family that might be having an impact.

The development of speech is a remarkable process involving the brain and the muscles of the face and throat, as well as the eyes. Children acquire language by watching how people communicate with them and trying to mimic the sounds they make. A baby first learns about speech by observing and copying the way a human mouth moves, and they will also see their face reflected in the eyes of the person addressing them. Speech then develops like a sensory dance of call-and-response between a child and their carer. Naturally, the degree of emotional security a child has gained through attachment to their carer influences how that process starts and continues. The technical word for this relational dance between carer and child is 'attunement'; as carers, we must learn how to do it.[1]

I imagined that one possibility in this case was that Besim was not speaking yet because something was amiss in his relationship with his mother. Not that she didn't love and care for him or was causing him any harm; a child's lack of language isn't always directly related to something that is happening to them. However, its delay can reflect some unresolved fear or distress in the people the child is closest to, and I understood why Owen might be concerned about Besim's mother.

'And the dad . . . ?' Owen shook his head, his face serious. 'Nadia was widowed in the war.' He added that up to now, she had been very cooperative and involved during the appointments with Besim, 'and they're always bang on time'. We both knew this was not a given in our trauma clinic, where patients' trepidation or a parent's reluctance to put their child through something difficult could sometimes mean they would come late or not show up at all. 'So far, I've had three sessions with them together. I've just started some play therapy with Besim – it's early days. But it's Nadia I'm worried about today, Gwen,' he said. 'She's quite young, just thirty.' Not that much younger than me, I thought.

Owen went on to explain that in their first session, Nadia had appeared eager and anxious to get some answers and try to help her son progress. When the child seemed a little wary at first of Owen's invitation to play with some toys scattered around the floor, she got down on the carpet with her son, kneeling and showing him the toys, trying to encourage him. But in the most recent sessions, as Besim silently began engaging more with Owen, making some little shapes out of clay and digging into a box of Lego, Owen became aware that Nadia had withdrawn. Then he realised she was weeping, so softly that he might not have noticed.

'It wasn't just brimming tears or something in the corner of the eye, Gwen,' Owen stressed. 'I'm talking poleaxed here; honestly, tears wetting the collar of her blouse.' It was a striking and sad image. I imagined it would have been difficult in the moment to judge how to react, with the boy there. 'What did you do?'

He'd decided not to call attention to what was happening to avoid distracting Besim. But he'd tilted his head and made eye contact to let her know he was aware of her tears. As the session ended, he saw she'd been able to pull herself together, wiping her eyes and managing to talk warmly to her son as she helped him put on his coat. Owen had quietly told her that he could see she was struggling. She'd insisted she was fine, just missing home that day. Owen had ventured, 'I think it may be helpful for you to speak to a colleague of mine. She works here in the building. Perhaps you could meet up, while I look after Besim?'

I understood his motive. Children's minds develop in the context of a secure attachment with the adults who care for them. This is the foundation of attachment theory, developed by British child psychiatrist and psychotherapist John Bowlby. He argued that early attunement with our carers creates a 'secure base' for our future relationships, especially at times when we are stressed or vulnerable. I was also reminded of a famous quote from the influential British paediatrician and psychotherapist Donald Winnicott, who said, 'There is no such thing as a baby.'[2] The idea sounds odd at first, until we realise that he is saying that babies do not exist in isolation; they can only grow in a relationship with someone who cares and thinks about them. So anyone who works with children must think about the health and well-being of their attachment figures as well, because that relationship affects the children's physical and emotional health.

Undoubtedly, helping Nadia could help Besim – but you can't make someone talk to a psychiatrist if they don't want to. 'What does she think about seeing me?' I asked. Owen gave a rueful shrug. 'She wasn't keen but said she'd do it if it would be good for her son. I doubt she's had any space to think about what's good for *her* for a long time.' She had already been in the UK for a few years and was not only a refugee but a widow, a young mother of a troubled son. How heavy a burden was that to carry? And yet, if I understood correctly, she had never accessed any mental health-care services until now, when her son needed us. How had she managed her distress all this time?

I agreed to meet her, and Owen made to leave, apologising for keeping me. 'Don't worry about it,' I told him, and I meant it. Any concern I had about being a little late home to see my own small son seemed insignificant next to Nadia's problems.

———

Before meeting Nadia, I read through Owen's notes, which were extensive and helpful. At their first meeting, Nadia had told Owen

('in proficient English') how Besim ate well, was developing normally, aside from his speech, and slept through the night. She had
proudly shown him the booklet in which the maternal health visitor had charted her son's steady growth. Owen noted that in each
session, Besim was non-verbal but responsive to her comments and
directions, which were mainly in English, not her native Bosnian.
That was interesting. Was it to ensure Owen understood, or was
it a way of separating them from their ravaged place of origin? It
might be both.

When the boy was out of earshot, Nadia had told Owen that
with her mother so far away, and not having been a parent before,
she had no idea what was 'normal', as if looking to him for reassurance. How well I knew that insecurity, the worry that my child
could be sick or suffering, and through some ignorance of mine, I
might fail him. Motherhood is so idealised that even the thought
of being a 'bad mother' can be intensely shaming.

By the age of two, Besim could not even come up with a word
for Nadia, which she said would be *Mama* in Bosnian. She tried to
repeat it to him while pointing at herself, but though he pressed
his lips together as if to copy her, no word emerged. How would
that feel for her or any mother? What does it signify if our child
doesn't identify us as *Mama* (or Mum, Ma or Mummy)? In his
written notes, Owen had quoted her saying, 'Besim sometimes
makes small sounds, like an animal . . . a dog . . . a squeak or a
whimper . . .', adding below that 'N says she doesn't always know
whether B is happy or in pain'. Owen had put a question mark in
square brackets beside the comment, presumably to remind himself to think about that more. I could guess why: if Nadia couldn't
'read' Besim's emotions, especially of pleasure or pain, that suggested a psychic distance that could be profound and damaging
for both of them.

It was a pretty brave admission, too. It's not always easy for a mother to talk about her fears and needs, let alone when she is being asked to share her vulnerability with a foreign professional whom she doesn't know. I felt a little hope: if she'd been willing to say this much, maybe she could, given some time, open up about the trauma of her bereavement and displacement from her home. But that was quite a leap, and there was no guarantee. I thought about her sitting in Owen's room, arms crossed and head bent, suppressing the sound of her crying to protect her son, making a considerable effort to 'hold it in'. I have lost count of the people who have told me that they fear putting their feelings into words in case they are overwhelmed by the emotions stored in their bodies and memories. At the same time, there comes a point when they may realise that if they aren't able to share or connect with others, they risk remaining isolated and dislocated, like in those science-fiction films where astronauts are lost in space, floating in endless darkness.

The first thing I noticed when I answered her soft knock at my door was her abundant hair. Her mane of black curls tumbled down almost to her waist, recalling an Italian film star. Her skin was pale in contrast, as if she rarely went outdoors, and her features were small and delicate. She wore a simple jumper and jeans, and the smoke-blue shadows under her eyes (much like mine) suggested too little sleep. She took a seat, clasping her hands in her lap. Rather stiffly, she said hello and introduced herself, and I did the same. I told her I gathered she spoke English very well and was sorry I did not speak her language.

I meant that literally and metaphorically. I was very aware that although we were close in age and both mothers of young sons, there was such a vast and crucial gulf between our experiences. I felt so ignorant of what Nadia had been through. Seeing your homeland become a war zone seems the epitome of insecurity and terror. I also had no idea at all of what it was to be persecuted because of your identity or to have to survive rejection and denigration by neighbours who used to be your friends. Those images from Bosnia of skeletal men and boys staring out from behind the wire of fenced concentration camps confused me by evoking another time and place. My brain insisted that they must belong to a different time in recent world history, incomprehensible today.

Nadia told me that she was a Muslim woman, born in Sarajevo, adding, matter-of-factly, 'My husband died in the war.' It seemed too soon to go deeper, but I asked when. 'Twenty-third of August 1995,' she answered immediately. The irony of that timing hit me with some force when I did some follow-up research: he had survived when so many young Muslim men had died in the preceding years, only to lose his life in the final fierce days of the conflict, possibly during the NATO bombardment of the capital.

Throughout this conversation, Nadia spoke softly and with composure, her delivery of the information bland. She seemed a long way from the tear-streaked, broken woman I'd expected to meet, based on Owen's account. Why was her presentation with me so different? Was it because we were both women? I reminded myself that people try and present well on first meeting doctors they don't know – and any refugee from a war zone may only have survived by putting up a good front. Or maybe she didn't trust me, given what she'd been through. Who could blame her?

She didn't wear a hijab over her hair. I figured that was probably true of many of her generation of Muslim women, who came of age in the secular, largely integrated European cities of the 1990s. When I asked about this, she sighed, then explained that the word 'hijab' isn't synonymous with headscarf, which I hadn't known. 'Hijab' translates from Arabic as 'curtain' and refers to a general covering for modesty, while there is a separate concept of a cloth that covers a woman's hair. I had the sense that people had quizzed her about this many times since she came to the UK, as if she were not a 'real' Muslim. She seemed to read my thoughts, telling me that some Home Office immigration officials had said something about her not wearing a hijab when she made her asylum application. Asylum seekers must demonstrate a 'well-founded fear of persecution', which can mean proving their religious affiliation as the reason for flight. I felt embarrassed to resemble even for a moment the bureaucrats who had questioned her cultural and religious identities. It didn't seem like we were off to a great start.

I changed the subject, asking her what work she had done in Bosnia, and when she told me she had been a teacher and an English tutor before the war, I said, 'Oh, that makes sense.' I added that she'd already taught me a few words and ideas I hadn't known; she must have been very good at her job. This raised the faint shadow of a smile. 'I taught English language, private lessons, in Sarajevo, to business people mostly. Before I married, I came here in 1990 for one year, to Leeds University. I studied modern English literature and poetry.'[3]

Until that moment, I had been considering whether she'd be better working with someone who shared her mother tongue. This is a vexed and intriguing issue in mental health. The eye may be a window into the soul, but language is a window into the human mind. Again, this is what makes us humans; our emotions, good and bad, generate our words. Moving to a new country and speaking another language (especially one that is new) impacts on how people express their feelings. As a therapist, I don't work with translators unless I absolutely must, because so much emotional communication involves idioms and metaphors that simply don't translate. In English, the list of metaphors to describe the inner experience is endless. Therapists often hear about the head, for example, as in 'I'm in over my head at work' or 'I'm head over heels in love' or (one of my favourites) 'It's doing my head in', where the 'it' is a painful emotion. But as Nadia had studied English – and not just the language, but our literature, including some of the poetry that I so love – I thought there was a good chance that we could establish a therapeutic alliance, hopefully strong enough for her to be able to trust me with what she had to say.

Around this time, I'd read a fascinating paper by the Indian psychoanalyst Salman Akhtar, who had moved from India to the US to work.[4] He noted that his experience of migration had parallels with the psychoanalytic concept of 'individuation': how people create a sense of self, first as a toddler and then in the psychic flux of puberty. Akhtar proposed that the first individuation for the immigrant occurs when a person starts to dream in their new language; the second when they begin to swear in that language; and the third starts when they realise they must inhabit a new identity. I found myself curious to know if Nadia dreamed or swore in English yet, but I didn't ask. I imagined these things would emerge in time.

I would start with the parts of her history that might be easiest to say out loud. We spoke about her arrival in the UK with Besim. Initially, they stayed with her old English professor in Leeds, with whom she'd kept in touch over the years and who had offered to host her if she decided to come to England. After a few months, she made her way to London, where she heard there was support for refugees and work to be found. She had begun her asylum application, but it was taking forever – a story I'd heard from many others. After staying in a hostel in Camden for nearly a year, she'd recently found a room in a shared flat and was making ends meet by cleaning houses with one of her girlfriends, who let her bring Besim with her while they worked.

I guessed she was not ready to discuss the war and what had precipitated her flight from Sarajevo. Instead, I asked if she could tell me a little about her life before the conflict, something about her childhood and family. Haltingly, she gave me dates and places and spoke a bit about her schooling, offering that she had one older sister but not much more. It was clear that from her point of view, this was just another interview by a professional who wanted information. I sought to get to something more personal. 'I'm so curious about what made you study English literature and what Leeds was like,' I prompted, thinking that could be a positive memory.

No luck – she seemed to become even flatter in her emotional presence. 'Leeds was very cold.' Her reply seemed to close the conversation down and even, I fancied, reduced the temperature in the room. I thought I should shift gear and talk more like a doctor. I asked about her general health (good), appetite (poor) and sleep (really not good). Nadia was consistently polite in her tone as she replied, but the context of what she said made me feel embarrassed again, especially when she commented that any health problems she had were likely related to years of poor diet and little sleep back home. I was reminded again that I knew nothing of living with war.

Throughout this exchange, there was no sign of tearfulness. 'Owen was concerned that you were crying and feeling sad, and maybe that's affecting your sleep now?' She nodded but said nothing. I took a moment, busying myself making some notes, as I considered what we should do next. There's an old convention in health care that one way to end a consultation is for the doctor to give the patient something – typically, a prescription for medication. This is especially common in psychiatry, where it takes time for people to get to know one another, and medication for low mood is an effective treatment for many people with depression. But I didn't know anything about Nadia's state of mind at this point. She could have some formal diagnosis, or she might be simply grappling with overwhelming loss, of not only her husband but her country, her family, her work – everything she had known about who she was. It would take some time to determine what I could offer to help her.

But there was one thing I couldn't leave for later. 'Nadia, do you ever get so sad that you feel your life isn't worth living? That you think about ending your life?' Immediately, she became animated, as if I'd pushed a button marked 'On'. The words burst out of her. 'No! No. Not ever. How could I do that to my son? Is this what you think? It is not so.' Her face was flushed pink. I responded calmly, trying to reduce the emotional charge in the air. 'No, it's not what I think, Nadia, it is what I must ask. Some people who describe the problems you have experienced may also feel suicidal. I just want to make sure I'm not missing anything.' She seemed to believe me, relaxing a little, so I thanked her for coming to see me and asked if she would like to talk with me again next time Besim had a session with Owen. There was a longish pause, and my heart sank. Had I botched our first meeting completely?

To my relief, she agreed and thanked me. I ventured another suggestion: 'Nadia, would you consider dropping in on our support group for people from the former Yugoslavia? We run one not far from where you live in north London.' She looked less sure about this but said she would try, if she could find someone to look after Besim. I wrote down the details, and she folded the bit of paper, carefully tucking it into her purse as she left.

Later, I talked to Owen and worked out with him when I would next see her. Before I went to catch my train home, I sent a message to Dragana, the Croatian colleague who ran the support group, giving her a heads-up that I hoped Nadia would be in touch. I wasn't sure if she would; it is part of our primate heritage to be anxious about joining new groups. At the time, I was aware that some therapists were voicing their scepticism about offering therapy, or any type of psychiatric treatment, to people fleeing war zones. These professionals[5] argued that it does not help people who have been brutalised by war to be seen and treated in their new host country as mentally unwell or labelled with terms like PTSD. With this in mind, the clinic had helped to develop some off-site groups for refugees, offering support rather than any diagnosis, and which were run by practitioners from their home country with a therapy or counselling background.

The groups were well-attended, and we received some positive feedback. People reported that they felt less like 'victims' and more like part of a community sharing a painful experience. The groups also offered a place to share information and, sometimes, food and drink from home. I knew some people valued attending because rather than talking about 'trauma', they shared practical concerns, like 'I can't get my parents to leave their house and join me here', or problems with immigration officials and paperwork. The group leaders also kept abreast of new resources to help them access basic needs.

I heard from Dragana that Nadia did not get in contact with her, but that was her decision. It did tell me that she could be avoiding communication with people from home, which might be significant. I was glad when she returned to see me, knocking tentatively on my door, then taking a quick step back when I opened it, as if she'd made a mistake in coming. Before she could change her mind, I held the door wide open and ushered her in. As she sat down, she thanked me politely for arranging for her to join the support group, but she hadn't felt comfortable about doing that. It wasn't right for her. Also, arranging childcare for the evenings that the group met was impossible. 'I understand,' I said. 'I really do.' I was curious about what wasn't right, and if that feeling meant she would reject my help, too.

———

But she did return to see me; we met a few more times over the next month. Mostly, we would talk about her son, not her – some little signs of progress she saw in him or her hope of getting him into a nursery school. There were days when we would just sit quietly together for a while, a familiar experience for me; sometimes the best service I can provide is to keep someone company, sitting with their thoughts and worries in silence. One day, I asked about her schooling. She had been a good student, she said, but rebellious in the way that younger daughters can be. Her sister wanted to stay home, marry and have a family, but Nadia wished to travel and interact with other cultures. She'd always been fascinated by all things English, listening to BBC World Service and studying English language and literature at school and university. She saw herself becoming a teacher, possibly even a writer. She'd seen an advertisement for a summer school in Leeds and had saved up the money to go. She had a wonderful two weeks immersed in modern English poetry – Auden, Spender, Eliot and Larkin. She dreamed of staying in England and applied to study literature for a year. She was introduced to Jane Austen, whose books she loved. I could identify with that; poetry and literature are how I understand the world. It occurred to me that her son's inability to speak would be even more upsetting for someone who valued words as much as Nadia did. I hoped we could talk more about that, but I felt too cautious to broach the subject yet.

She said she liked watching BBC programmes to improve her vocabulary and pronunciation. The first time we ever laughed together was talking about the television adaptation of *Pride and Prejudice*, and the famous scene of Colin Firth emerging damply from the lake. After that, I felt bolder. I said, 'Nadia, we've never spoken about Besim's father. Is it possible for you to say something about him, even just how you met him? What was his name?'

Nadia visibly stiffened; her face was expressionless. 'I do not think I can say.'

'Ah, that's important. Because . . . ?'

'Dr Gwen . . .' She had recently begun to address me like this, which I'd taken as a good sign. Staring at the floor, she began to speak, her voice low and halting. 'I – I have never spoke his name since he died. I cannot. When I went away from Bosnia with Besim, I just had to . . . not say. In this way, I could survive. Go forward. I must find a place to stay, do everything and take care of Besim and myself. Alone. And if I did not ever say his name, then I could do those things, I knew.' She looked up at me, dark eyes meeting mine. 'I know it.'

'What do you know?' I asked her gently.

'If I speak of him, I think I will . . .' She paused, then brought the palms of her hands together, hard, the cracking sound like a gunshot in the small space, making me jump a little. Then she gritted her teeth, as if to stop any dangerous words from escaping, eking out only this much: 'If I say his name, I will explode.' Then she dropped her head into her hands, covering her face momentarily, her anguish palpable. I waited as long as it took until she could look up at me again. Her eyes were still dry but rimmed with red, full of pain. 'Do you see, Dr Gwen? Then who will be here for Besim?'

This may sound illogical, but from a psychotherapist's point of view, her emotional rationale was faultless. Nadia feared disintegrating under the weight of the painful emotions she had held in check all this time. This was the first sign of a connection between her son's muteness and her fear of speaking about her husband or even uttering his name. I said something to that effect, and she looked stricken. 'Do I hurt Besim for not speaking? If I die, then I hurt him much more.' I assured her I understood her concern. 'It sounds very hard. You fear you will' – I searched for the right words – 'perhaps let something terrible out?'

There was a silence. Then Nadia whispered, 'What can I do?' Far from certain myself, I offered, 'Could you try, safe in here with me, to say Besim's father's name?' Nadia shook her head vehemently, and silence fell again. We must have sat there for several minutes, our heads bent as if in some shared prayer. Then she made a noise, like a grunt or cough. I saw that those sad, dark eyes were wet for the first time, and she was trembling, evidently on the edge of letting go at last. She made another noise, a hiccup becoming some sort of word.

'Ivan.'

Then, drawing out the two syllables, 'I-van.'

As if surfacing from a long time underwater, she gasped for air, her face contorted, and I thought that surely she would weep now. Yet she didn't. I let myself repeat his name aloud: 'Ivan.' I was honouring her effort and contemplating with sorrow this young man who would never grow old, the father that Besim would never know unless his name was spoken.

Nadia made a movement, her eyes going to the wall clock. There was a little time before the end of our session, and I had to make sure that she would be all right to leave after this important disclosure. 'Is there anything you need right now, Nadia? How about some water?' She declined but then took the offered glass. 'Nadia, you've taken such a risk today and been very brave. It's nearly time for us to stop, and I need to know if you're okay with going home now. Is there anything you want to say?'

She shook her head. I thought about her words, her fear that she would explode. 'Are you in pain?'

She gave the slightest nod. I knew better than to move or ask for more and waited. Eventually, she sighed, reaching round to fumble for her coat on the back of the chair. 'I feel shocked, I think . . . but I feel something also . . . I can't say. A small correction – no, I mean connection to him. But I cannot do more today.' She pushed herself to her feet, and I stood with her in the small space, eye to eye.

'I understand. But I'd like you to call me tomorrow to tell me how you are. Will you do that for me? I think it's essential for you and for Besim. And is it okay if I tell Owen about our session today?'

'Okay. Tell Owen. And I will call.' She ducked her head and all but ran away. Watching her go, I wished she had more support at home. This session had been huge for her. I thought of Ivan again and all the grieving war widows whose men vanish, their names inscribed on silent monuments.

———

I was mightily relieved when Nadia rang me the next day, as promised, and told me that she had got some sleep; not much, but some. She said she was only now aware of how tired she felt. We planned to meet again in a few days, and then I went looking for Owen. I told him my story. What did he make of Nadia's long-held fear of speaking her dead husband's name? What he had to say was fascinating: 'Now it makes sense.'

He went on to explain that although Besim was still not verbalising, there was some progress to report. After weeks of playing with toys, the boy had reached for the big crate of crayons in the corner and was beginning to make drawings. Typically for a child of his age, these were impressionistic: a wiggly blue line for the sky, a green puddle of grass below and a few scribbly stick figures and shapes to represent animals or people. At Owen's suggestion, during last week's session Besim began to draw pictures of his home, the flat he and his mother shared with some friends.

'What are they like?' I asked. 'Colourful chaos, really.' Owen smiled. Using different crayons, little Besim had made several drawings that mainly featured wobbly geometric structures with tiny figures inside, Owen told me. 'And something else . . .' At this point, he broke off, going to his filing cabinet to extract some papers. 'Here you go, have a look at this one.' I frowned, peering down at it as Owen waited, expectant. I didn't know what I was supposed to think or say. I couldn't have identified the crayoned scrawls on the page as a flat or even a building. As a parent, I've been gifted my share of brightly coloured drawings over the years, proudly brought home to stick on the fridge (often accompanied by a conversation along the lines of: 'Oh, is that a big monster?' 'No, it's you, Mummy!'). But child psychotherapists are trained to get alongside their patients, helping them use images and shapes to learn about their minds and feelings and search out sense without words.

'Look here, Gwen.' Owen pointed to the inside of one rectangle, where Besim had drawn two tiny cross-shaped sticks, one topped with a mass of black scribbles. 'Besim and his mum. I checked with him to make sure that was her.' Was that Nadia's curly black hair or some impression he had of tangled turmoil inside her head? I didn't make that guess aloud, aware of my ignorance in this specialised field.

'What's this dotted line down here, do you think?' I asked, peering at the space below the figure of Nadia, where what appeared to be a mini-parade of ants spilled down in a vertical line. Owen leaned over my shoulder. 'She's crying.' I exhaled softly. 'Oh.' Our children see everything, take in everything. You do not need to be a child psychologist to know this. I had experienced it as a mother, and in my training I was reading about how children zero in on their parents' faces, taking in every emotional cue we give them, including ones from our own childhoods, those 'ghosts in the nursery', unbidden visitors from the parents' pasts.[6] I knew enough to understand that the drawing was not a photograph; Besim may not necessarily have seen his mum crying in their home. Owen suggested to me that her son was 'drawing attention' to her grief and could also be manifesting his own tumultuous emotions. I noticed, too, that both stick figures had little circles for eyes – and no mouths.

'What if he doesn't want to talk because he's angry?' I mused. Owen looked quizzical. 'I could be wildly off-beam here, but what if he's angry because he thinks his father abandoned them?' Even as I said this, I realised I was being too literal, ascribing adult thoughts and feelings to a little boy.

'I don't think this is anger, Gwen.' Owen turned away to rummage through the stack of drawings and pulled out another one. 'Take a look at this.' One big, lopsided rectangle, drawn in red crayon, took up much of the page. Two of the tiny, scrawled stick figures I'd seen before stood close together in the bottom right-hand corner – no ants this time. But off to the left was a balled mass of black wax, the paper appearing slightly torn – as if the boy had urgently tried to erase or cover up something he'd drawn. I held the page up to the light and thought I could detect an upside-down 'V' of stick legs beneath Besim's intense scribble, conveying a sense of some other person in his home space who was 'cancelled' or hidden beneath a black cloud of feelings.

'I still don't know what happened to Ivan,' I ventured. 'And Besim couldn't have seen and remembered his death, could he? He would only have been a few months old.' Owen frowned. 'Yes . . . but I have an idea Besim thinks he knows something but doesn't know what it is, other than it seems too scary for words.'

I thought of this little boy and his mother huddled together in their box. 'They've been everything to one another for so long,' I said. 'Clearly, she does all she can for him. But this silence, with her terror of speaking about his father . . . maybe it's left him with the belief that it's dangerous to speak. I'm thinking that if she can take a risk, then maybe he can, too. How much does her state of mind affect his?'

'That's the point, Gwen. We know that children are closely tuned in to their mothers' minds. It's possible Besim has understood that speaking is dangerous, if his mummy is frightened to talk about certain things.' Owen looked at me, his gaze serious. 'This reminds me of the traumatic grief we see in a case of domestic homicide, where children lose both parents after one partner kills the other.' That brought me up short, recalling some recent work I'd been doing in Broadmoor with a group of men who'd killed members of their families, several of whom struggled even to name their offence. At best, they might muster something like 'my index' (the offence they were sentenced for). Bearing witness to a violent death, even if you caused it, can leave you at a loss for words.

I also realised that because I didn't know what had happened to her husband, I hadn't even asked Nadia about nightmares and flashbacks, the phenomena we sometimes see after a traumatic death. What would it take to help her tell me that story and describe what she and possibly her baby son had witnessed? The images in Besim's drawings gave me something I could take back to her. They were a communication from child to mother: he was aware of her sorrow and of something hidden or masked.

———

I was unsure what to expect when Nadia and I sat down for our next session. What came across was anger, not directed at me but nonetheless palpable in the room. I began by asking some general questions about what had been going on for her since we last spoke. She sat there, picking at a loose thread in her jumper, offering nothing. I could see she was biting her lip as if to stop anything escaping, and I searched for the right words to help her get past that. Looking back now, I can see how much I still had to learn about speaking the language of therapy, which is less about solving distress than listening to it, sitting with discomfort and finding gentle ways to enable the patient to go deeper and reveal more.

I talked a little about Besim's drawings, but she seemed dismissive. 'Children's pictures – they don't mean anything.' I knew I was on the wrong track and decided now was not the time to focus on Besim. Instead, I asked her how she'd felt after our last session and what she'd said about her husband. I did not use his name. Her head jerked up; her voice was sharp. 'It is over. I spoke his name. I did not die, yes. Life is this for me now. Being alone, no husband.' She lifted her hands and spread them out, palms up, as if to say, 'What else is there?'

'Yet you and Besim are together.' I had to say it.

Nadia snorted, her anger hot. 'They killed his father, and I can do nothing. Bombs come down and kill everyone, my friends, my cousins, they kill and kill. I protect my son, but for how long?' I could only validate her fury: 'Yes. So much killing . . . too much to comprehend.' There was a long pause during which nothing happened. We sat together, two women silenced by the useless violence and cruelty of war. Then, with a quick movement, Nadia tugged at her sleeve, thrusting her wrist towards me to show me her watch. I'd not noticed it before, though she may have been wearing it all along. 'It is his. All I have.' I held my breath, and soon enough she added, 'And Besim.'

The watch was nothing special – too big for her slim wrist, with a worn brown leather band and a glass face with Roman numerals. Then I realised the time was wrong, out by several hours. She quickly covered it with her sleeve. 'It stopped when he did – I will not repair.'

A line formed in my mind, so apposite it escaped my mouth before I could check myself. Not quite under my breath, I quoted that tragic opening line of Auden's poem about a much-loved man: 'Stop all the clocks.'[7] I wasn't sure Nadia had heard me, but to my surprise, she quickly chimed in, reciting, 'Cut off the telephone. Prevent the dog from barking.' I knew she had studied English poetry, but I still felt amazed and touched by this connection. I worried that I might be the one to cry now.

We've all been faced with grieving people in our lives, and it is natural to be affected, but generally, my professional training has encouraged me to contain my emotions in front of patients – this is their time. As doctors, if we get overwhelmed by our feelings, we may be less able to help and heal others, so we learn to create a distance between our reactions and the work. I reflect on them later in supervision, where I meet with a colleague to discuss my interactions with patients. There is a downside to this, however: it is possible to be too distant and come to see a patient's painful emotions as 'symptoms', not communications.

I suppose I have found it is easier to maintain boundaries in a forensic setting because the offenders I work with have long histories of mental illness and have acted violently. In trauma clinics, people are often much like me, leading quite ordinary lives, until X or Y happens; there is more resonance than dissonance between us. Sitting with Nadia that day, I struggled. Her story of the watch and recital of the poem were unbearably moving, but I didn't have the confidence to know if I should let my sorrow show. Today, I would probably allow myself to cry; I have done this with patients at times, if I was sure the tears were about their story, not mine. I have learned this can be another sort of witnessing – and a communication that I have heard and seen their sorrow.

I had no idea where the session would go next. Nadia spoke first. 'We study Auden in class. In Leeds.' Of course. I took some breaths, steadying myself. Then, as I watched, her lower lip began to quiver. Rocking a little in her chair, she moved her head from side to side, as if reprimanding herself or in battle with a rising tide of feeling. What could I say to help her? So much longing, so much loss; what Shakespeare called the grief that 'whispers the o'er-fraught heart and bids it break'.[8] What could I say to relieve her? All I could think of was to ask, 'What if you just cried, Nadia?' At that, she squeezed her eyes tightly shut in an effort, it seemed, to prevent her pain from pouring out of her limbic system and into the space between us. She grimaced, her mouth opening a little, then closing again. I thought she was trying to speak, so I leaned my head close to hers: 'Tell me.' She barely breathed the words into my ear: 'I am afraid I will never stop.'

That human fear of the undammed heart, I knew it well. But what I have seen and what I know is that, in time, the tears will come to an end. That day, Nadia began to cry and cry, and it went on for the rest of our session. The tears continued when she returned for the next one, and the one after, too. We would talk a little about Ivan, about her different memories of him, until one day, she was ready to tell me how he died. It seemed to come up easily, despite all her previous reticence and withdrawal. They had this tiny new baby, and they were hungry, she began. They needed bread, and Ivan had gone to try and find some in the bombed-out city outside their apartment. 'I had to eat. Besim was so small, and I made so little milk for him . . .' It was early in the morning, she said. Ivan promised he would not be long. He said he could smell the bread from the bakery in the next street; it was ready. Before anyone else was awake, he would run to get what he could for them.

'I held Besim in my arms, and when Ivan was leaving, he stopped to kiss us each, one' – she mimed a kiss – 'then two' – another kiss in the air. 'Here.' She pointed to her forehead, between her eyes. 'It would be ten minutes, he promised. He can run fast.' As she spoke, her left hand automatically covered the watch face on her right wrist, and her face coloured. I had an image of a handsome young man, pausing with his hand on the door, naturally wanting to bestow parting kisses on his beloved young wife and the baby son she was cradling. And none of the trio had any awareness that it would be their final communication, their last contact. Outside, the familiar sounds of gunfire and muffled explosions mixed with sirens heralded the new day. 'Ten minutes,' Ivan said, and he was gone.

Five minutes became ten. Then twenty became an hour. Nadia's voice fell to a whisper, and now she wrapped her arms around her torso, rocking backwards and forwards from the waist, as if to soothe herself. 'I think I knew . . . I have a chair my father made when we married. It moves, rocking. I sit in that chair with the baby, always far from the window in case the bombs break it. I wait and I wait. I am for hours just sitting there, looking at the wall. I try to feed Besim, but there is no milk in me, so I rock him to sleep, he is good boy.' She closed her eyes, rocking, remembering, so much pain in her face. 'There is noise, loud noises in the street below, some shouts. But all I hear is the clock tick. I know. I know. The baby sleeps. I do not cry. I listen for footsteps on the stair . . . and when his brother opens the door, I say nothing. I do not cry. I know.'

Nadia continued with her memories of that dreadful day. She took the baby to Ivan's parents, and then she went with her brother-in-law to find her husband's body. It was in a makeshift morgue set up outside the bombed-out bakery. Rows of bodies, some under cloth. She spoke of the smell of burned bread, the screams of women, mothers bent double in the debris. Someone showed her a watch, and she confirmed it was Ivan's. They indicated a black plastic bag at her feet. It didn't seem big enough to hold a great, tall man, her Ivan. That was all that was left of him, they said. His body had been obliterated. At this awful point in the story, she cried and cried, and as she wept, I saw how her thumb rubbed the watch face, moving back and forth on the glass as if to clean it – clearly a habit and a comfort to her. I thought it was remarkable in the circumstances that she had been able to identify Ivan this way. Stopped time, hands frozen at the last moment of his life. No wonder it was so precious to her.

After twenty minutes or so, she reached for what was left in the box of tissues on my desk, dabbing at her eyes, moaning a little as if in some physical pain. When she had composed herself a little, I asked, 'What's happening for you now, Nadia?' She swallowed.

'I have a headache. I feel tired.' 'Grieving is exhausting,' I said from the heart. It is not just emotional labour; it is bodily labour, and causes neurological and endocrine responses that may give rise to hypervigilance, sleeplessness and changes in the immune system.[9] C. S. Lewis observed that grief can feel like being concussed 'or mildly drunk'.[10] It can leave us so raw that we sense that even being touched by a feather would hurt, so we disconnect and retreat from others. I've also had people tell me that when not at its most acute, the grief of loss can almost be boring – but in the sense that it bores into you, inescapably, that constant drilling into the heart wearing your spirit thin and rendering your inner landscape devoid of colour, endlessly grey.

Nadia was staring out of the window, across the rooftops, towards another place and time. 'That bag . . .' She was still with Ivan in her memory, and though our hour was up, I let her talk a little longer. Her words tumbled out. 'So awful . . . a black bag. Like for rubbish. I cry out when I see it in the night. Besim is close to me in our bed, touching my hair. I think I must be waking him if I shout in the dream. And I tell him it is nothing, go to sleep. But he knows. He knows.' She shook her head at the impossibility of keeping her distress from her son. 'I think this is in the picture you told me of. He is feeling my memory. I do not know how, but he is.'

I had nothing to say to this. Perhaps she was right.

———

The war had ended soon after Ivan was killed, and in her next session, she told me how she'd come to leave Sarajevo. 'No one wanted me to go – Ivan's family, my family – but I had to get away. Besim could not grow well on the ground where his father was killed. I wanted to go away to England, a place where we are safe, and I was happy once. I think, I will speak another language and never talk of Ivan . . . and not cry. Never cry.'

I felt I could finally ask her what had shifted, what had happened that day in the room with Owen when she had first let her tears flow. She spoke in the present tense, the memory essential and live for her. 'When Owen kneels to talk to Besim, he is smiling, very kind, and I see Besim looking up at him, watching. He has Ivan's eyes, eyes that show all he feels. Besim is so like his father. I watch them play and I cannot stop the tears, they fall out.' The little ants on the child's drawing, tumbling down the page. Again, I thought of Lewis on grief, trying to tell himself 'Love is not the whole of a man's life' and 'People get over these things', until there is a 'sudden jab of red-hot memory and all this "commonsense" vanishes like an ant in the mouth of a furnace'.[11]

There was more crying in later sessions, with both me and Owen. Nadia continued to find watching Owen playing with Besim especially painful; she told me about one session where they began to build something with Lego, and how much it hurt knowing that Ivan would never be able to do this with his son. Nadia pressed the middle of her chest with her palm, describing how her heart felt 'so tight'. A tight heart was not a metaphor I'd heard before, and maybe it was a literal translation of her idiom, but it made absolute sense. Consider how excruciating it can be to wear something too tight, or our English idiom 'being in a tight spot', which denotes the impossibility of a situation. I had also heard or read about old pocket watches that broke when they were too tightly wound. Perhaps this is what Shakespeare meant about the import of giving sorrow words: a heart wound too tight may break.[12]

I noticed after some months that Nadia stopped apologising for crying. It is an odd thing, but people quite commonly say 'I'm sorry' as they start to weep. Are they worried that their sadness will be contagious? I was concerned for her when she cried, but at the same time, I was glad because she had learned to trust me with her grief. And she had so much of it. In truth, I think I also found it exhausting sometimes.

Later, I reflected on how Nadia had coped since Ivan's death; not just coped, but survived remarkably well, far from home and family, caring for a young child, securing food and shelter, navigating paperwork and bureaucracy. As she sought help, she'd had to contain her grief, because if she'd allowed herself to experience it any sooner, it might have overpowered her and inhibited her ability to survive in a new country. Putting feelings on hold in this way, the 'numbness' that so many trauma survivors describe, is recognised as an early part of the body's traumatic stress response. Such deactivation may be a protective mechanism that stops people from becoming overwhelmed by their feelings, so that they can take action that will save themselves and others, too. Think of all the stories of soldiers in the thick of battle who have got themselves to safety while carrying a wounded buddy, despite their own grave injuries – something echoed in Tom the POW's nightmares. Often, they will recall nothing of this later and are amazed to be told of it; it is as if the mind protectively deactivated to allow them to emerge from the battle alive.

Even in peacetime situations there are individuals who switch off their emotions in the face of a traumatic event. However, the study of emotional regulation suggests that prolonged deactivation can be harmful because it interferes with the processing of feelings, which helps an experience move from the present into the past.

We had needed to help Nadia say Ivan's name, and she would need to let Besim hear it, too. In time, I hoped the boy would learn from her how much she loved his father, that he had eyes like his, and that one of Ivan's last actions was to kiss him. He would be fortunate to grow up understanding that crying can be a type of freedom; the vertical line of ants in his crayoned drawing moving out of frame might be heading somewhere new.

Owen and I met to discuss what next for this mother and child. We knew Nadia and Besim had used up most of the sessions usually allocated to people referred to the trauma clinic. If we did not restrict the number, we would be unable to help more people. Under pressure to save money, the new Labour government was instituting a raft of 'efficiency reforms' intended to modernise the NHS, and the administrators wanted evidence of *throughput*.[13]

I felt reluctant to bring the work to an end, but Owen was more positive. He reported that Besim was smiling more when they met, even laughing as he played – a much less solemn child than in earlier sessions. He had also begun to say a few words, mainly confined to naming animals in English: 'cat', 'doggy', 'horsey', 'cow'. Owen had given him a talking book: Besim would press the button and hear 'meow', 'woof', 'neigh', 'moooo'. The boy loved it, copying each one, repeating them over and over. He was not so scared of his own voice now; maybe he felt others were ready to hear what he had to say. Owen thought he was connecting, and the attachment between Besim and Nadia seemed more secure. The boy was interacting more playfully with his mother now, and he would turn to look at her when he moved towards Owen, checking her gaze, attuned to her in a way he hadn't been before.

I thought Nadia could now be a good candidate for a bereavement support group, maybe one conducted in her own language. There, she could explore what was important to her religiously or culturally, maybe even decide what saying goodbye to Ivan could look like. I thought about my own experience of funerals and burials, cemetery visits and memorials, and all the ceremony and gathering we have in our culture when someone is lost to us. It occurred to me how often the Auden poem Nadia and I had talked about, 'Funeral Blues', had been read at the services I'd attended, and I considered the import of standing up at these events and speaking of the person lost, whether in your own words or the language of others.

————

Nadia and Besim's story opened my eyes to grief as a trauma and how a death without warning affects people of different ages in different ways across time. Before meeting her, I had imagined Nadia as fragile, a lost soul rendered helpless by losing her husband. The word 'bereave' comes from an old English word meaning 'robbed'; one terrible summer morning, Nadia had had her identity as Ivan's wife snatched from her, leaving her exposed and shocked. On arrival in England, she'd had no choice but to don the new, ill-fitting garments of other identities, including 'widow', 'asylum seeker' and 'refugee'.

But she was still a mother and had to attend to her bond with her growing son, who looked at her with Ivan's eyes. Besim had witnessed his mother's sorrow but also felt her terror of it. She once told me she feared it would cause her to 'explode' – she would fragment into non-existence, as her husband had. Instead, the invisible blanket of her grief had covered her and her son, rendering Besim silent. The parent–child attachment so crucial to language development was mending now. I had witnessed how this woman's love for her son had prompted her to bring together the speechless, grieving mind and the bilingual, articulate but avoidant mind. In doing so, she had begun to grow a new identity, which made room for her sorrow. She moved into that third individuation.

I could not help but think of T. S. Eliot's 'Here I am, in the middle way . . . each venture is a new beginning / a raid on the inarticulate'.[14] I was beginning to discover more about what I could and could not do with language as I searched for the way I wanted to be in the world as a psychiatrist and psychotherapist. I hoped to use the best words I could, in the best order, as the poets do.

The Lucky Ones

On the day of the crash, weather conditions were ideal: *temperature 17°C, SE wind at 6 knots and no cloud cover.* As we hear from the crew every time we get on a plane, safety is paramount, and it was chilling to me, a frequent flier, to read the catastrophic error summed up in the report with the modest phrase *inadequate pre-flight engine inspection.* Someone had missed a fault: something complex to do with wiring, which I admit I skimmed over. Pilot error had been excluded as a causal factor.

This was my introduction to the work that lay ahead of me. The legal firm representing three survivors of a helicopter crash had instructed me to assess its impact on their psychological health. The air accident report was the first of many documents in 'the bundle', a pile of legal paperwork much like the other expert reports I'd seen during my forensic training. The author's language was carefully bland and short on sensory detail, which I was glad of when it came to the moment of disaster. The helicopter's engine had cut out about an hour into the flight. The black box later confirmed that the pilot and co-pilot had identified a field a short distance away for an emergency landing, but they were losing altitude too fast. Instead, they smashed into the side of a hill. Four tonnes of metal met granite, the cockpit splitting from the passenger cabin on impact. The report described how the tail rotor sheared off, *with the acute tilt causing Passenger D (female, 47) to be propelled through the open door into the rotor blades, resulting in decapitation and dismemberment.*

The pilot and co-pilot (*males, 41 and 32*) were also killed on impact, their bodies crushed inside the cockpit. Passengers A,

B and C (*female, 29, males, 47 and 24*) were thrown clear when the rest of the helicopter tumbled down to the foot of the slope. To my astonishment, the accident investigators reported that all three of them *sustained only minor bodily injuries*. Nothing was said about any psychological injuries. That was going to be up to me.

This assessment came relatively early in my career. While I had done a lot of medico-legal work in the criminal courts, assessing people for trial and so forth, I had less experience in cases like this, involving claims for damages in the civil court, allegedly caused by traumatic events and requiring diagnosis, which is rarely straight-forward.[1] Although I'd willingly taken the case, I felt slightly nervous. Today, I suspect lawyers would look for an expert with more of a track record than I had. But I was working in a specialist clinic for trauma then, and Wendy, the solicitor for the three survivors, thought that was important. She was looking for someone competent in assessing adults with a range of mental health problems that might be linked to trauma, and that was something I could do. My previous forensic work for the courts was also a bonus because I was familiar with the problem of translating psychological issues into legal concepts.

———

Wendy was twenty years my senior, and her manner was as authoritative and professional as her navy-blue suit. She suggested that I meet with her clients at their workplace, which might be less stressful for them, and she volunteered to drive me there so that she could update me on the progress of their case. As we set off, I sat back, ready to listen, feeling a bit like I was back at school.

'My clients are claiming damages against the parent company behind the helicopter charter firm that owned the helicopter that

crashed. They're one of the big five in private aviation, you know?'
I didn't know a thing about the aviation world, but I said nothing;
it was enough that I understood that a lot of money was at stake.
As part of the claim for any psychological damages, Wendy went
on, Passengers A, B and C would need to show that they had suffered some post-traumatic psychological injury, and the evidence
for that would come from my assessment.

'In claims like this, the company usually denies any fault, and
the plaintiffs must prove the company's negligence caused harm
to the claimants. The insurance firms these companies use never
want to pay out, so they resist such claims vigorously. But the
good news in this case is that the company has no choice but to
admit liability. So now we just argue about how much the survivors' compensation should be.'

'I'm amazed they survived at all,' I said, grateful to get a word in.

Wendy nodded, keeping her eyes on the road ahead as she skilfully manoeuvred her car through the morning traffic. 'A miracle,
really. Which may be why the company is keen to settle this claim;
it's probably cheaper for them that way. My clients were just very
fortunate. Relatively speaking, of course.'

'And all three have returned to work since the accident, you
said?'

'They took some sick leave for bumps and bruises and the initial shock, but yes, they've been back for some time. You may want
to ask them about that.'

I was puzzled. 'I will, for sure . . . Doesn't going back to work
impact their claim, though? I thought people got damages for
time off work and lost income?'

'True enough, in many cases,' Wendy said. 'But they also get
compensation for any mental health problem that's caused by the
crash, even temporarily. We also need you to comment on the

prognosis – not just in the short term, but into the future, especially regarding their ability to work.'

Commenting on prognosis is not as easy as it sounds, and I said so. So many people, especially those who deal in facts and logic, seem to think that doctors, including psychiatrists, can make accurate predictions on how people will recover after illness or what quality of life they will have. If only. Psychological recovery after trauma is complex and depends on a range of factors, including what else is happening in your life in the aftermath of traumatic events; life doesn't stop while you're recovering from trauma. Ironically, legal proceedings for recompense can be a further source of stress that can slow recovery or even prevent it.

Wendy carried on with her briefing, oblivious to my worries. 'In general, those claimants with the worst post-traumatic stress symptoms will get the maximum award, but the law says that people's resilience shouldn't be held against them or reduce the size of their claim. Just because they're strong doesn't make the company less liable, right?' She cocked her head to one side, expecting me to concur, and I did. She made perfect sense.

I had a question she might not like: 'Presumably, to help their claim, people tend to talk up their problems, maybe downplay their strengths?' She glanced over at me but did not seem to take this personally. 'Most people don't. Look, Gwen, we don't want a hired gun who will say what we want them to say in court. We need the best independent psychiatric assessment that you can do, giving careful consideration to how these people's lives have been affected by this awful accident. They may *seem* okay, but are they? That's the question.'

We were arriving at our destination: the headquarters and factory of a large kitchen manufacturer. As we pulled in, I vaguely recognised the brand name emblazoned above the tall gates.

'Anything else I can tell you?' Wendy was asking as she parked the car. 'Will I have to give evidence?' I blurted. I hadn't done that before in a civil court. She touched my arm, her voice reassuring. 'I doubt it will come to that. As I said, they've indicated that the insurers want to settle rather than go to court.'

———

The room we'd been allocated for the interviews was in a small, prefabricated cabin, one of several dotted around a large campus of low-slung modern constructions flanked by a few Victorian-era red-brick buildings. Wendy said she'd spoken to all her clients and explained the plan for the day. Each would see me for a couple of hours, and then she'd talk to them about the case. She had arranged for me to start with Lisa (Passenger A), then Martin (Passenger B) and finally Caleb (Passenger C). 'Two before lunch and one after, okay?' Before I could answer, she went off to fetch Lisa.

I surveyed the drab room. A metal desk and two chairs sat on worn industrial carpet tiles, and a plain clock ticked loudly on the wall. I was glad that at least there was a window to one side looking out onto a bit of green space, with a few trees providing shade. I sat down, pulling paperwork out of my bag. Rather than just making the three of them go through their statements with me, I planned to explore issues that hadn't come up and try to understand the differences in their experiences. I'd read some recent research about 'hot spots', the moments of peak emotional distress in people's recall of a trauma.[2] I would certainly be asking about what seemed to be the worst, or 'hottest', aspect of the accident for each of them. I heard voices outside and jumped up to move the chair facing mine at an angle, so that whoever I was meeting could look out at the view rather than me if they began to feel uncomfortable or upset. I thought I would want that option if I were in their place.

There was a sharp rap on the door, and I turned to see Wendy ushering in a tall, athletic-looking woman of about my age, dark hair drawn back into a high ponytail, her blazer and skirt appropriate to an office job. 'Lisa,' she said, holding her hand out to me, her eyes meeting mine. Wendy was off, saying, 'Leave you to it. Back in a couple of hours.' I thought Lisa looked surprised that it might take that long, but she settled easily into her seat, with none of the nerves I might have anticipated. I first checked that she understood why we were meeting: I'd be writing a report others would see, her lawyers included, so our interview was not confidential. She cut me off. 'Wendy went over all that. It's fine.' I sensed this woman was keen to get this interview over with, and I couldn't blame her.

Could we start by talking a bit about what had happened before she got into the helicopter on the day of the crash? 'Sure,' Lisa said. She'd been the first to arrive at the airfield that morning. She recalled it was one of those lovely March days, cool but clear, spring in the air. Her words were already painting a better picture for me than the accident report's terse summary of wind speed and median temperature. Dev, her 'then-boyfriend', had come to see her off, and she'd got him to photograph her in front of the helicopter with the camera she'd given him for Christmas. Then he pulled the strap over his head and hung it around her neck. 'You're the one with the view,' he'd told her. That camera was never found, she added, her voice steady, matter-of-fact. 'Must have been smashed to bits.'

I let that vehement idea of destruction sit in the silence between us for a moment, before I asked why she was in the helicopter in the first place. There were some bare facts in the legal bundle referencing an office outing, but I wasn't clear on all the details. 'It was a prize,' Lisa explained. Their annual company Christmas party always culminated in a charity raffle – 'Everyone got tickets.' This year, there were spa days and theatre outings, and for the grand finale, four winners would get to take a chartered helicopter flight. 'Actually, I was gone before they pulled my ticket out of the drum,' she explained. 'The music was too loud, everyone was getting pissed . . . I was getting in my car when my friend Sue came rushing out to tell me I'd won. I couldn't believe it. I never win anything, you know?'

She told me that her colleagues Martin and Terri had soon joined her by the hangar that morning; Caleb, whom she didn't know, was late. On the tarmac, a few workers busied themselves preparing the helicopter, which was shining silver in the sun. 'Looked like a big metal insect,' Lisa said. 'I remember thinking it seemed flimsy up close, compared to an aeroplane.' She did not sound scared as she said this, but her language suggested some fear. Martin was quite forthcoming about his pre-flight nerves: he'd taken a little nip of brandy in his coffee that morning, he'd admitted to Lisa and Terri, sotto voce.

Lisa had a particular memory of Terri: 'She was smiley, excited. Pulled out a floral Thermos from a matching tote bag and waved it at Martin, saying, "Just good old lapsang souchong for me! Can't wait!"' Lisa was a good mimic, her accent turning Midlands, voice rising an octave, as she conjured a woman whose earnest enthusiasm matched her appetite for a nice cuppa. 'Did you know Terri before? Were you work friends?' I asked. 'Terri? Um, not well. She was older than me, and she was the boss's PA. She was a good soul, though. Always kind. Poor woman.' Lisa stopped speaking, and I felt her emotional discomfort. It occurred to me that she might have some guilt about surviving, but I didn't want to assume anything at this point, thinking I'd return to it later.

They were about to board when Caleb arrived. In his statement, he'd said that he overslept that day and 'almost missed out'. 'He turned up just in time for a speech about safety procedures,' Lisa said, 'and then the pilot asked if we had any questions.' I remembered Martin's statement had also mentioned this detail; he had asked what seemed to me an eminently sensible question: would they get parachutes? No, came the reply, helicopters don't tend to carry them, partly because they fly at such low altitudes. The pilot would make an emergency landing in the 'unlikely event of a problem'.

All their statements tallied in describing how once the helicopter lifted off, the noise of the engine and the rotor blades was overwhelming, too loud for conversation. The pilot and co-pilot, separated from them by a glass panel, wore headphones and communicated through a crackly speaker to the cabin, telling them the flight plan and pointing out some landmarks along the way. As this was a tour, they flew with an open door. Lisa told me that she remembered seeing the university where Dev worked, and she leaned past Terri, who was nearest to the door, snapping multiple pictures as she tried to get a good shot.

It was only when they moved north-west, away from the city, flying across a green and brown patchwork of farms and fields, that Lisa first thought she heard 'something funny'. They were maybe an hour into the flight. She'd been on aeroplanes before and was aware there were always weird mechanical noises; it was probably nothing, she told herself. But Terri must have heard it, too, because she leaned in close, her mouth right next to Lisa's ear, shouting, 'What was that?' a bit hysterically, as Lisa recalled. 'Like she'd never flown in her life. Maybe she hadn't.' Terri's anxiety was soon mirrored by activity in the cockpit, with a flurry of radio communications and busy hands tapping at the complex instrument panel. I knew from the accident report that the black box recording indicated that the pilot had reported a problem and then confirmed his coordinates to air traffic control and his intent to make an emergency landing. The black box recording ended with 'Oh shit'. Nothing more. I would later learn that this is often the last thing pilots say before impact.

All three survivors said their memory of what happened next was blurry. Time speeded up; time slowed down. The rapid loss of altitude and the gut-wrenching lurch and tilt of the helicopter were so disorientating that none of them had a straightforward sequential narrative. Someone screamed, maybe Terri. Lisa used an accounting word: Terri was suddenly 'subtracted' from their midst, four becoming three. Lisa said something about the 'massive bang' of the impact and rolling down, 'tumbling over and over'. My mind had them like tickets in a whirling raffle drum, their bodies thrown around the metal capsule yet somehow miraculously *only sustaining minor bodily injuries*. They each described becoming aware of being alive but feeling numb, almost as if they were in a dream, which is typical of the first phase of psychological survival after a traumatic event. Nobody mentioned what they'd seen of Terri's awful demise, and I wondered if or when they might do so.

The accident had happened about a year before our interviews, so I knew that the three survivors would have worked through the initial phases of the post-traumatic stress responses that affect most people, such as the intrusive recall of events, which can come at any time, and hypervigilance about possible threats that remind them of the trauma. It takes time for people to make sense of their experiences and create memories, and for most people, these symptoms resolve over the next six to twelve months. After a year, it's only a minority who may still have the classic indicators of post-traumatic stress disorder. Today, there are multiple ways of diagnosing PTSD, but back then, our approach was pretty simple. I would begin by giving everyone a form listing a range of symptoms and ask them to tick those they had experienced.[3] I would also be on the lookout for other psychological signs of distress, such as relational problems at home and substance misuse.

I told Lisa about the questionnaire and asked if she felt able to fill it in for me.

'You mean right now?'

'If you would.'

She took a pen and began ticking boxes, adding in the comments section how 'for weeks' she'd jumped at the sound of aircraft overhead and felt reluctant to leave the house. She slept poorly, 'for the first time in my life', and had 'a few' nightmares, although after a while, those had stopped. All of this had resolved over the course of the last year. Her doctor had given her some medication to help her sleep, and she'd had time off work when needed. As I read through the form, I asked her if she'd felt supported by family and friends during that time. Everyone was great, she assured me; her sisters and parents came to visit often, and Dev sat up with her at night when the nightmares were bad.

'Anything else I can tell you?' Lisa seemed keen to end our interview, glancing at the clock as if she needed to be elsewhere. I said we wouldn't be much longer but asked if we might talk briefly about her life before the accident.

'What do you mean?'

'Like where you grew up? How you came to work here? Stuff like that?' She seemed puzzled, until I explained that it would be helpful for my report to know a little more about her background. She didn't resist and seemed to relax a little as we moved away from the topic of the accident. She'd grown up in Peckham, in south London, the eldest of three girls. Her mum and dad ran a corner shop; the five of them had lived in cramped quarters upstairs. Money was tight. 'But we were happy as a family,' she said. 'Us girls were close to Mum and Dad and each other. Still are.'

I asked if any other adults had been involved in her upbringing, and she spoke lovingly of her grandparents, who had lived nearby and would take care of the girls if Mum and Dad were busy. In later years, as a teenager, Lisa would look after the younger ones, making sure they did their homework and had their tea. 'Quite the little mother, I was.' She smiled. 'Do you have any children now?' I asked. She shook her head. 'Will do, I hope – Dev and I just got engaged. Last month.' I noticed the ring then, sparkling on her left hand. 'What brought you to this place, this line of work?' She shrugged. 'Good at maths, really. I went to uni in Birmingham and thought I might teach, but in the end, there were some decent jobs in the accounts department going here. I applied, and they hired me, so . . . here I am.' I made a note of her positivity, her openness and apparent willingness to engage with my questions.

'This is a good place to work then?' She confirmed it was. 'Even though the company's massive, hundreds of employees, it feels like a family.' I thought there might be a connection between her security at home and in the workplace, but there was no time to go into that now. I needed to explore any history of trauma prior to the accident. As an adult, had she ever had any frightening experiences before the crash, including things like assaults or harassment at work or home? I got an emphatic denial. And as a child, had she ever been abused or harmed by anyone? She looked as if I'd asked whether she'd ever had two heads. 'Me? God no. My parents weren't like that. I mean, Dad might give you the odd slap if you stayed out too late or back-chatted Mum . . . but no – no.' I was interested in her switch to the second person, as if she'd distanced herself from receipt of 'the odd slap', and I made a note. She frowned. 'What's that got to do with the case or anything? Do you put all this into your report?' I told her that was unlikely – without adding that, in truth, I was more interested in how she talked about her childhood than the actual content of her memories.

———

It was around the time of this case, while I was still qualifying as a psychotherapist, that I became increasingly interested in how childhood attachment relationships might be protective in terms of surviving trauma. This was an old idea, but there were some new ways of assessing attachment in adults, one of which I was being trained to use: a specialist tool called the Adult Attachment Interview (AAI), in which people answer a series of twenty open-ended questions about their childhood memories of being cared for. Areas covered include factual details, like whether someone experienced bereavement or parental divorce in childhood, before moving on to more subjective memories. So we might ask for some adjectives to describe a relationship with a parent, and whether they felt closer to one of them in early childhood. The interview is taped and transcribed, and people's attachment security is coded, or 'scored', for things like coherence, lack of recall, vagueness or idealisation of parents.[4] Through this process, a great deal can be learned about someone's attachment 'style', within four broad categories: secure, avoidant, anxious or preoccupied, and disorganised. Given the complexities of the human experience, this categorisation isn't cut and dried, and the primary value of the interview is in understanding more about how people think and feel based on language and narrative – and what that reveals about how they deal with stress.

I was still learning about the AAI and wasn't proficient in its use. I couldn't be sure, but judging from how she spoke and the content of her memories, I thought Lisa probably had a secure attachment in childhood, which might explain her apparent resilience after this dreadful experience. In most communities, across nearly all countries and cultures, security of attachment is the norm for about 60 per cent of people. Only later would I wonder whether she had been a little avoidant, given her brisk manner, her way of closing down conversation and, ultimately, how little she gave away.

———

We were coming to the end of the interview, and I checked if she had anything she wished to add, something that might have occurred to her since she had made the legal statement or had come up while she was completing the questionnaire. She shook her head. 'Can't think of anything else, Doctor. Sorry.' Another closing down of conversation.

'No need to apologise . . . that's great. I do have just one last question.' Her eyes were darting again to the clock on the wall. 'Um, okay. Go ahead.' 'Lisa, what was the worst thing about that day?'

'The worst thing?' I had her attention. 'Let me think.' My mind conjured images of dismembered body parts. Then she said, 'I guess it was the silence.'

'The silence,' I repeated. There was nothing silent about a helicopter crashing into a mountain. What was she talking about? As if she could read my thoughts, she added, 'After, I mean.'

She shifted in her chair to look away from me, out of the window, her eyes on the empty sky. It was a minute or so before she spoke again, her voice quiet. 'We were there for ages. Hours and hours. As soon as I realised I was alive and all, I looked around, and Martin was there . . . right over there . . .' She gestured towards the green area outside the window, some ten yards away from where we were sitting. I felt she was watching a film playing out as she described the memory.

'I crawled over to him, not knowing if he was dead or what. And then I remember I was so, so glad when he lifted his head and looked at me. I didn't see anyone else at first. It was just the two of us, breathing, looking at each other, and there was no noise at all.'

When she stopped, I waited as long as possible before prompting her. I noticed that she was using the past tense, which suggested that as difficult as this memory was, her brain had 'filed it' in the past.

'And then?'

'Then I saw something in the grass a few feet away. I thought it was a tree branch or something. Turned out it was . . .' She swallowed hard. 'It was her leg. And then we saw an arm, a bloody stump . . . and suddenly there was Caleb, running towards us, scrambling through the grass, his face all scratched and bloody, eyes wide, hair sticking up. He looked about five years old. Threw himself down next to me in the dirt and started to cry, but not making any sound, just these horrid, gasping sobs, his shoulders shaking . . . I noticed he had some teeth missing, and I thought he'd fallen on his face and knocked them out. After a bit, I reached over and took his hand. And still nobody said a word. No birds sang. There was nothing, just this long, long silence.'

The accident report included an account from first responders, saying that the trio had been found just as she described, huddled together in shock, about ninety minutes after the crash. But in Lisa's mind, time had stretched and distorted, which is not unusual in a single-event trauma like hers. 'Hours and hours,' Lisa said again, her voice dropping to a whisper, as if a certain hush was essential to capture the experience. 'The silence felt, I don't know . . . smothering . . . but we couldn't break it. None of us said a word.' She paused, took a breath, sat up straight and reverted to her previous tone. 'That's it, Doctor . . . that was the worst part.'

I was so wrapped up in her story of lost time that I'd lost track of how long we'd been talking. Wendy would be back any moment. I thanked Lisa for what I knew must have been a hard conversation and warned her she might have more thoughts about the crash in the next few days. If it felt difficult, did she have people she could talk to? 'Sure,' she said, getting to her feet. 'Is that it?' As she turned the door handle, she stopped, turning back to say, 'That was better than I thought it would be, you know.' I thanked her for that welcome feedback, and she left, her back straight, head high, ponytail swaying a little as she walked away.

I had a quick look at the questionnaire she'd filled in, which indicated she was unlikely to meet the criteria for a PTSD diagnosis as it was defined at that time. (The criteria continue to evolve, as part of the broader and unending debate about 'what counts' as trauma and mental disorder.[5]) I thought she was one of the fortunate majority of people who surmount the natural distress that comes with disastrous events. While the helicopter crash and its aftermath were still alive in Lisa's mind, that did not define her life or stop her from working or relating to others. I remember thinking her engagement ring was also a good indication that she was moving forward. In the words of one trauma expert, people don't have to 'get over' their trauma, but they need to 'get on with it'.[6] So it seemed with Lisa that day.

———

I went through my notes, preparing for the next meeting, and soon Wendy was at the door with a middle-aged man in tow, announcing as I stood up that Martin was feeling a bit worried about the interview, 'Aren't you, Martin?' She was clearly hoping I would have the right words of reassurance. I tried to meet his eye, but he looked down at his feet. 'How can I help you with this?' I asked.

'Do we have to do it in here?' His voice was low and shaky. The bare little office suddenly felt crowded with three of us in it. Trauma can induce claustrophobia in some people, and I reminded myself that a helicopter cabin is not big. I made a quick decision. 'It's a nice day. Shall we sit outside somewhere and talk?' Behind his back, Wendy beamed her approval, giving me a thumbs-up before she left.

Everything about this man seemed slow, from his speech to his gait. We found a large bench under one of the chestnut trees in the green space outside and sat down. Martin did not make eye contact with me, resting his forearms on his thighs, hands clenched together, head bent. His clothing, complexion and hair were dull and colourless. He looked every one of his forty-seven years, and then some, with deep lines etched around his eyes and mouth. Some reddening of his nose and a puffiness about his face hinted that he might be drinking more than was good for him. Sometimes a diagnosis leaps out at you before you've asked someone a single question, and I was reasonably confident this man was depressed, struggling in a way that Lisa was not.

I carried out much the same interview with him as I had with her, but Martin's responses were halting, and his voice lacked energy, although I could see he was trying to cooperate as best he could. He became more animated when he talked about his past, almost like it was another country, a world away from this dark time. Born in Bolton, he was the only child of older parents, 'Spoilt and doted on by my mum.' But the marriage broke up when he was in his teens, and he and his mother moved down to the Midlands. He didn't see his father often because 'he got a new family'. I made a note of that; although parental divorce and separation from a parent are not uncommon for children, such an early loss can be linked with later vulnerability to depression.

Martin and his mother settled near where we were meeting, in a suburb of Birmingham. He married Laura, a local girl whose father worked at the kitchen company. Through that connection, he got his job as a joiner. It sounded like things had gone well for them over the years: he spoke warmly of their three children – 'all doing well' – and repeatedly praised his wife. 'She's always been a rock for me, especially since—' He broke off then, as if unwilling to refer to the reason for our meeting. 'Since the accident?' I prompted. He glanced up at me briefly. 'That's right.'

Martin explained that things had already been tough for him in the months before the crash. In the autumn, his mum had been taken ill and died quite quickly, within weeks of a diagnosis of advanced bowel cancer. This hit him hard, and he began worrying about his health. 'I, er, had some tummy troubles,' he said awkwardly – passing blood, nagging abdominal pain. His GP sent him for a load of tests that weren't conclusive, but Martin became convinced he was going to die, just like his mum had. Losing her also left him feeling 'alone in the world', like an orphan. 'Your father isn't alive, then?' I asked. 'He's long gone.' Martin's voice was curt, slamming a door on that painful part of his life.

His anxiety about his health got worse, and shortly before the office Christmas party, he'd seen a doctor, who put him on some pills – he could never remember the name, started with a 'C' . . . 'Antidepressants?' I prompted. 'Yeah.' He still wasn't sitting upright or looking directly at me, suggesting he might be a man who felt ashamed of seeming vulnerable, of needing treatment for his pain.

He'd never even wanted to attend the office Christmas do in the first place, he said, but Laura liked dressing up and going out. She'd bought them a job lot of raffle tickets, but he hadn't thought they'd win anything – 'If you want to call that a "win".' It was plain he didn't want to talk about the accident yet, so I decided to return to it later.

'How are things at home now?' I asked. Martin swallowed hard. 'Um . . . the house feels small. It's like I'm trapped in there.' 'Trapped?' I mused. 'Maybe you don't like confined spaces since the accident? Could that be why you preferred to talk with me out here?' He acknowledged I could be right.

'What do you do when the house feels small, Martin?'

'I go out a lot, walking around for hours sometimes . . . In all weathers, all hours, I don't care, I just have to escape.' He sighed. 'And I know it's hard on Laura and the kids. She thinks we need a change of scene – wants us to move away from here, maybe down to the coast, if we ever get some money from this case. Put it all behind us. As if.'

I pushed a little then. 'Would you like to move away, Martin?'

He finally turned, angling his head to face me, his eyes weary. He was going to tell me something important now, I thought, 'She's happy, I'm happy. I don't really care. I can get work any-where. Nothing seems that important now.'

This was the language of clinical depression. I needed to explore some things in his statement, but I thought this might be a big ask today. 'You sound exhausted,' I said, wanting to convey my sympathy. He slumped, head down, and muttered, 'All the time.' I felt fatigued just listening to him. This interview was becoming an ordeal. I decided to offer him some relief.

'I'm not going to keep you much longer, Martin.'

'We're done?' His voice sounded hopeful.

Almost, I told him. If he could just answer one more thing for me. He immediately looked worried, but mumbled, 'Go on then.' I put to him the same question I'd asked Lisa at the end: what was the worst thing about the crash that he remembered?

'The worst?' His voice deepened, like he was being dragged down by the weight of the traumatic memory. He shifted position on the bench and then twisted his watch on his wrist, as if these small movements might buy him some time. I watched and waited. After a bit, he shut his eyes and began to speak.

'I think . . . it was when the engine cut out, and we started to fall, and maybe it happened fast, like seconds, not minutes, but to me, it seemed like we went into slow motion, like going down a mine shaft or something, falling, falling, falling, and there was no bottom, and that poor woman—' His voice broke. He was not going to articulate whatever thought or memory had arisen about Terri's death, but I could feel his tension, a palpable sense of the awful. He sniffed, then wiped his nose with the back of his hand. 'It was like I had a cold stone inside me, Doctor – the fear. I was sure that this was the end, that I was dead. And it still feels that way now. Even if we were the lucky ones.' He'd said it, not me. Combining the words 'dead' and 'lucky' sounded odd and ominous. I had to ask the question that psychiatrists must put to anyone who's depressed: 'Martin, do you ever feel so bad you think life's not worth living?'

He shook his head. 'I couldn't do that to them . . . to the family . . . not after all this.' I asked him how he managed his bad feelings, other than by going on walks. As I'd thought, he'd taken to drinking more than he should. 'It helps, it numbs me out at least. But they don't like it much – and they let me know.' I could imagine the friction, the rows. His family had been through a lot with him, and their concern and upset about his drinking could only deepen his feelings of shame and isolation. Despite his denial, this man was a genuine suicide risk, I thought.

On the other hand, I was heartened that he'd had the strength to get back to work and was at least open to planning for a future with his wife and children. It was also good to know that he'd sought professional help in the past for his anxiety and depression – and that he'd been able to tell me as much as he had. 'Martin, I'm going to make a plan with Wendy to see you again another time, all right? That's enough for today.' He was happy to be released, but when he shook my hand and thanked me, I felt some warmth in his manner.

———

Wendy and I discussed the morning's interviews over lunch in a corner of the vast company cafeteria. I told her that I thought Lisa seemed to have survived quite well; she didn't meet the criteria for any current diagnosis, although she had experienced some PTSD symptoms in the past. I shared my concern about Martin's mental state, surmising that his experience of bereavement and a health scare shortly before the crash meant he was already in a vulnerable state when helicopter met mountain. Already preoccupied with death and dying before the accident, he'd then nearly been killed and had witnessed the violent end of Terri and the pilots at close quarters. I wondered if he had such a focus on death that he might take his own life. I would need to call his GP to tell him what I thought about Martin's suicide risk and suggest he might need more vigorous treatment. Wendy pushed her tray aside. 'That's no problem, Gwen . . . Anyone's treatment needs are hugely relevant to the claim. Now, shall we meet Caleb?'

Caleb was in the machine shop on his break, and Wendy had arranged for me to talk with him there since he refused to come and see me in the office. 'He's also feeling anxious about this interview,' she explained. I assured her it was okay for me, as long as we could be alone, and she said everyone had downed tools for lunch, so she hoped I could at least make a start. 'He's a bit agitated,' she warned. As if to demonstrate this, we found Caleb hammering a metal sheet on a workbench in a big, deserted warehouse space filled with lumber and machine parts. Down at the far end of the narrow, hangar-like room, I spied a couple of men sitting together, eating from lunchboxes and playing a game of cards, but they were well out of earshot.

Wendy made the introductions, and I said hello, but her client did not acknowledge me. 'Caleb, I know this isn't easy,' she said, stepping towards him, 'but we just discussed this, and you agreed to see her.'

Her voice was kind but firm, and her head girl quality was evident. Caleb shoved the metal sheet aside, and it clattered to the concrete floor. He turned and faced me. 'All right then . . . let's get this over with.' His voice was rough and unwelcoming, and I felt uneasy and irritated. I would have to try and establish a rapport as best I could and try to understand why he was so prickly. 'Can we sit and talk here?' I indicated the workbench. 'Suit yourself,' he said. Wendy left us, and Caleb looked at me with narrowed eyes while I dragged a stool over and sat opposite him, placing my notes and pen on the bench.

Once I was settled, I studied him for a moment. I knew he was in his mid-twenties, but it would be easy to mistake him for a secondary-school student playing truant. He had what might be called a baby face, with little or no facial hair, and he was short in stature, shorter than me. His jeans sagged below his hip bones, and he was so thin and pale I feared he was malnourished. When he spoke, I glimpsed the missing teeth Lisa had mentioned; there were at least three or four dark gaps.

'What am I meant to say to you then?' he asked abruptly.

'Where would you like to start?'

'She said to tell you how I'm feeling so you can write it down or something.' A London voice, his '-ing' sounding more like '-ink'. I also thought he might have a slight speech impediment related to the missing teeth. The hostility in his language could be a defence against feeling frightened, and I needed to do what I could to defuse that. As with the other two, I explained my purpose, adding, 'I'm not here to upset you. I'd just like to ask about how you've been since the accident.'

He made a dismissive gesture. 'Can't you just put down that I'm feeling like shit? Why do I have to explain anything to you? *They* should explain how they let us go up in that chopper.' His voice rose, and one of the men down the far end raised his head, looking across at us briefly. Caleb kicked the metal on the floor with the toe of his trainer, causing a cymbal shimmer to echo around the open space.

Big, big fear there – and considerable anger, too. I suspected it flared often, probably ignited by almost any interaction; I was just his lightning conductor for the day. Even if he wasn't angry with me, I could see how this interview might feel like an intrusive attack, at a time when he was feeling vulnerable. I'd seen this in my forensic work with other young men who felt trapped and exposed.

Validating his feelings was the first step, I knew. 'This process is awful, Caleb, I agree. Why don't you just say what you can? I've got questions, but we don't have to get into them all today.' I was already thinking he was too aroused and agitated to be doing this, but I wanted to allow him to settle so that he could get it over with, if possible. He gestured at my notes and pen. 'Write this down, Doc. I want the helicopter company to pay up. And maybe' – the sweep of his arm took in the work area around him – '. . . this lot, too. Some raffle prize, right? Company should have vetted them properly.' It was the first time I'd heard negativity directed towards the employer; neither Lisa nor Martin had shown any. Caleb then dismissed the notion of the helicopter having a mechanical failure, saying he knew a thing or two about engines. The problem, in his view, was that the pilot was 'crap' and undoubtedly caused the crash because he hadn't a clue what he was doing. I knew otherwise from the accident report but did not argue.

'Did they tell you I had a breakdown after?' he said suddenly, lifting his chin, crossing wiry arms over his chest. I thought for a moment about how to answer. 'Wendy mentioned you'd been struggling.' He glared. 'Struggling?' His voice was sarcastic. 'That what you call it?' He shook his head at my stupidity. 'I saw that woman—' And there he faltered. I didn't try to help him, willing him to go on, to express the fear and rage I could see roiling in him. 'You can't imagine. I'm lying there, mouth full of dirt, don't know if I'm dead or alive, or if this is a bad dream or what, and then I saw it next to me and it—' He stopped again.

'It?'

'Yeah, her hand. Nails painted pink and all. Jesus.' His face was reddening, his jaw set.

I decided to move on to what came after the crash, noting that he'd returned to work after just a short period off. 'I need the job,' he said bitterly. 'If I keep getting signed off sick, they'll fire me. I can tell the company wants to get rid of me anyhow – they all think I should've died instead of her. They'll probably try to push me out, but I'll sue them if they do.'

His rage and paranoia were growing, pulsing, radiating from his body, and I was beginning to feel uncomfortable, a sensation I'd learned to take seriously in my forensic training. I should end the interview immediately. Nothing I could say would make him feel calmer or less afraid.

Just as I was thinking this, he confirmed it. 'Shrinks . . . fucking shrinks and lawyers!' His eyes were fierce, and flecks of saliva formed at the edge of his lips. Then he spat a crude insult at me, adding, 'You're all the same, you lot . . .' Instinctively, I drew back, flinching a little, which caused me to almost fall off the stool. I had to rebalance myself quickly, taking a moment or two to gain my equilibrium. That's when I saw his discarded hammer lying on the floor, the one he'd been using on the metal sheet earlier. I realised his eyes were on it, too. There was a pregnant pause – that frozen unit of time that precedes a burst of action. Not, thankfully, him assaulting me with a hammer. Instead, he kicked it, sending it skittering across the floor, shouting at me, 'I've had enough!' before stalking off.

The men at the other end of the warehouse jumped up in alarm, and I prayed they would stay where they were. I was worried about what Caleb could do if someone more threatening than me were to confront him now. Luckily, the next person he ran into was Wendy, waiting just outside the door. She skilfully talked him down and took him home, before collecting me and making the drive back to London. En route, we both agreed he needed urgent help. He presented a risk to himself and others and shouldn't be in a workplace where many tools could be used as weapons.

The next day, Wendy rang to tell me that he'd been admitted to a psychiatric hospital, detained involuntarily and diagnosed with paranoid psychosis. I had witnessed the start of this mental collapse in the factory workshop. By the time the admitting psychiatrist saw him, he had progressed from being antsy and angry to being frankly disorganised in his mind. I knew this could happen in a matter of hours and was not surprised when Wendy reported that he was 'in bits' – highly agitated, wild of eye and insisting that 'people want me dead'. I probably wouldn't be able to see him again for some time.

I told Wendy I didn't think I needed to meet with Lisa again, but while we waited to hear how Caleb was faring, I asked if I could see Martin for a follow-up, maybe in a month or two, once he'd seen his doctor. She was happy to organise this, and we agreed to do it at her office in London. As soon as Martin arrived, I noticed a difference in him; his face and demeanour suggested he was more present and alive. He shook my hand when it was offered, then dropped into the seat opposite mine, immediately making eye contact. 'How are you?' I asked, and before the words were quite out of my mouth, he answered, 'Better.'

Martin's GP had changed his meds for depression and got him some cognitive behavioural therapy, which had made a big difference. 'I've packed the job in, Doctor . . . I think it's for the best. We're moving down to Kent – found a place near Deal.' So Laura had prevailed. I mentally applauded her. 'The company wanted to give me a leaving party, but . . . you know.' He shrugged. 'A leaving party?' I prompted. Martin shook his head. 'Told them I didn't fancy it. You understand why.' I did. I could see he was still a saddened man, much altered by his experience, but when we parted, I felt heartened that he seemed more engaged in the business of life. When he completed my PTSD questionnaire, he did not report any current symptoms.

———

Before I followed up with Caleb, I was able to request and review his GP's notes to see if he'd had any mental health problems before the crash. As with most young men, his file was slim. He'd enjoyed good physical health thus far, or at least he had not consulted his doctor often. But he'd been known to the child protection services, and some reports referred to him being taken into care briefly as a boy. I also saw a couple of assessments by child psychiatrists who had seen him years earlier – presumably, the first of his 'fucking shrinks'. On the face of it, Caleb was someone whose attachment security had been disrupted as a kid, which I thought might explain his extreme reaction to the crash. He was also the only one out of the three who, according to Wendy, was not in a long-term relationship at the time of the accident, nor – to the best of her knowledge – had he ever been. That could be important. Good intimate connections in adult life may allow people who have insecure attachments in childhood to grow a secure base, which can be a protective factor should they experience trauma.

Eventually, I heard from Wendy that Caleb was about to be discharged from the hospital and was ready to meet with me again to complete our assessment. When I arrived at the psychiatric facility where he'd spent the last few months, I noticed immediately how, as with Martin, his physical appearance had changed. He'd put on weight – a side effect of the anti-psychosis drugs – and it suited him. His attitude was also entirely different: he welcomed me with a smile and a handshake, almost like this was our first meeting. I had to check if he even remembered our encounter at the factory, but he assured me that he did, then swiftly apologised for his sudden exit that day: 'I just had to get away, Doctor.' I told him I knew he'd been dealing with a lot, and he'd done what was right for him at the time. I wasn't surprised to find him so altered; it's not unusual for acute psychoses to resolve just as quickly as they start, especially if people are treated with medication and get appropriate care. None of this meant that Caleb had left the trauma behind; he might just be able to speak about it now.

We went on to have a civil and thoughtful conversation, far better than I had anticipated. As we strolled together in the hospital gardens, I was struck by how friendly Caleb was with the other patients and staff, greeting people by name as we passed, which seemed positive and the opposite of paranoid. I noticed how he tensed up when I asked if we could talk about the crash, and I reassured him that he could say as much or as little as he liked. He nodded, saying he knew it had to be done for his legal case.

We sat on a bench, and I let him tell me what he remembered about that day. 'Where shall I start?' he asked. I smiled encouragement. 'Anywhere you like.' I noticed that he spoke in the present tense, using phrasing like 'I'm late, nearly missed the flight' and 'All of a sudden, the engine cuts out . . .', as if the experience was still live for him. But this time, he seemed more sad than angry about it, and when I asked him about his worst moment, he spoke again about seeing poor Terri's severed hand, and how he would never forget it.

'What about now?'

'The worst thing now?' He scratched his head. 'My mum.'

'Your mum?' I hadn't expected that.

'I'm worried for her.'

I wasn't sure what this had to do with the crash, but it was my job to follow where he led. He started to tell me a little about his early life. His mother had run away from her home in Dublin when she became pregnant at sixteen, finding work in a pub in

London. She often dated men she met there, he said, adding, 'She was always begging me to call them Dad, and I knew they couldn't be, plus I was scared of them, and of her, too.'

I kept my voice neutral. 'What scared you about her?'

'Whatever set her off. She'd go on the piss and scream at me, saying I was just like my father, which was confusing since I didn't know who that was . . . And then her boyfriends beat her up sometimes. In front of me.' His was a too-familiar unhappy tale. Not only had I seen other boys like him who'd grown into troubled, alienated men, but I'd also met mothers like his who felt helpless and hopeless, unable to protect their children and overwhelmed by feelings of powerlessness, shame and guilt.

'Did your mother ever hit you?' I asked, and he glanced away. 'Not hard.' I thought of the sad little boy, Willie, in *Goodnight Mister Tom*, who would insist that his mother was kinder than most, giving only 'soft beatings'.[7] What about her boyfriends? Had they ever hit him? 'Too right,' he laughed, without mirth. 'See here?' He pointed to the gaps in his teeth. 'I'd only just grown in my big teeth when one of them did that. That was the first time I ran away from home. I did that so many times I landed up in care.' When his mum eventually found safe accommodation away from the pub, his care order was lifted, and by the age of fourteen, he could start living with her again. 'So we moved up here to get a fresh start, you know?' He found an apprenticeship that would ultimately lead to his job at the kitchen factory. 'I only started last November. Hired just in time to go along to the Christmas raffle, wasn't I? Lucky me.'

I asked if he thought he'd be ready to return to work at the company soon. 'I don't know . . . I was in a bad way.' He'd decided to follow his doctor's suggestion to continue with rehabilitation and allow people to take care of him a little longer. 'They're good here.' I felt I could now give him the PTSD symptom check-list I'd asked the others to complete. In the 'past' column, he ticked 'intrusive images' – that severed hand had been persistent, although it was now largely gone. He also put two emphatic ticks for irritability and anxiety in that column and one tick for each under 'present'. I think he was about to do the same for anger when he stopped, pen raised. 'I was so angry before. Not so much now, but back then . . .'

'Go on,' I said, unsurprised.

He told me how he blamed everyone for the crash, not just the pilot but his employers and his manager, and spoke about how furious he felt that nobody had protected him. Recalling how he'd gone at that metal sheet during our first meeting, as if hammering the forces that oppressed him, I couldn't help but think of the violent boyfriends and his mother's inability to protect him as a child.

He finished the form, scribbled his signature and handed it back to me. 'But I've got something else that isn't on your list.'

'What's that?'

'Well . . . I still don't feel right.'

'How so?'

'I'm like . . . I feel like a snail without a shell,' he said. 'And I get these dreams.'

'Nightmares?' He hadn't ticked that symptom on the PTSD questionnaire, but by this time I knew that not everyone gets them.

'Not exactly. More like memories. I'm little, and Mum sends me to my room. "Be a good boy and go to sleep," she says, but I hate being alone. Then I hear things – shouting, thumps. I'm crying and screaming, bashing on the door, but nobody comes.' We had stopped by some flower beds, seed labels marking out what was planted, and I stood with him there in silence, letting him take the necessary time. He cleared his throat.

'It's not like that now. Mum's been good to me since the crash, and even before that . . . But back in the day, she was . . . well, the dreams are bad.' He turned to look at me, his face contorted. 'She locked me in my room all day sometimes, Doctor. I was a kid – couldn't have been more than six or seven. I think I went mad in there, screaming, thinking she'd never come back. So now I worry that if I go home, all this stuff will come boiling out of me. Like, at her. What do you think I should do?'

I was seeing him as an independent expert, not a therapist, and I had to be careful not to step outside my role. I did not doubt how much he needed to talk about his childhood and explore his past fear of his mother. Now that he was more mentally stable, I understood that it could seem like 'the worst thing' to want to strike back at her for past neglect. I thought their troubled attachment history could also explain why he'd experienced me as threatening that day in the warehouse, when I'd had an intuition that the hammer had crossed his mind as a weapon to use against me.

None of this was for me to raise with him directly. I told him that I thought he was right to consider staying at the facility and having more therapy, looking at his childhood fears as well as the crash. I didn't add that I had plenty of evidence that he had suffered a significant psychological injury because of the accident, affecting his day-to-day functioning, especially his work. I thought that when the court came to allocate damages, Caleb would get more than Martin or Lisa, and I hoped he would use that money to continue his recovery after being discharged from the hospital. I would highlight his need for long-term therapy in my report.

When I finished my assessments, I sent them to Wendy so she could discuss them with her clients. Since I'd seen both men twice, I suggested I meet with Lisa again after all. Wendy said she'd ask her, then came back and told me her client was okay with what I'd written in the report and didn't see the need to meet me once more. My work was submitted into evidence.

Over a year later, there was, as predicted, a settlement, which, to my great relief, meant that I would not have to go to court to defend my opinions. As I had guessed, Caleb got the most significant amount, Martin a similar sum, and Lisa somewhat less. That seemed a bit unfair to me, with Lisa's resilience meaning she was compensated less than the others. I thought back to the original accident report, which cited their *minor bodily injuries* and offered no indication of the significant impact the crash had had on their minds. I marvelled once more at how common it is in our culture to treat the body and the mind as completely separate entities.

These three crash survivors could be considered fortunate, despite their ordeal. Martin had said it: they were the lucky ones. They were alive. But I was learning that this might be all they had in common; three people could have almost the same traumatic experience, but with three very different survival reactions. It was plain that emotional stressors and lack of childhood attachment security are significant risk factors that make a difference in how resilient you might be after experiencing major trauma. Lisa seemed the 'luckiest' in that regard; that's why I thought she had suffered but also recovered, striding positively into the future. She seemed to have left her awful experience behind. Today, I know better than to think that trauma stories always end so neatly.

———

About three years later, I received a call from a psychiatrist in Birmingham who ran a perinatal service looking after new mothers with mental health problems. She told me she was treating

Lisa for acute post-partum psychosis, a severe mental disorder that has been recognised since the time of Hippocrates.

My colleague knew about the helicopter crash and had gained permission for us to speak about Lisa's case. She wanted to know how Lisa seemed back when I met her. Puzzled and concerned, I shared with her what I remembered, and my view that Lisa had no active mental illness when I assessed her. She was very unwell now, my colleague reported, ever since the birth of her daughter six months ago. 'Severe depression, believing that she and the baby are about to die, even hallucinating that their bodies are decomposing. Her husband thought she was going to bury the baby alive.' I struggled to reconcile this presentation with the Lisa I'd met. Something macabre was bursting out of her memory and manifesting as a paranoid belief that both she and her baby were dying or dead.

Today, I know that about one in every thousand women in the UK who give birth develop post-partum psychosis – a relatively consistent figure globally. I've learned about this because I assess women with this condition for the family court; they have become so anxious and suicidal that they can't look after their children. Sadly, a very small percentage of mothers with post-partum psychosis are also at increased risk of killing their own children, and in those cases, my forensic work coincides with the work of perinatal psychiatrists. I have been privileged to learn from some wonderful colleagues in the field of maternal mental health that this condition is, thankfully, very treatable, if recognised.[8]

Lisa's psychiatrist and I did discuss whether her post-natal psychosis might be related to a rare psychiatric disorder called Cotard's, or 'walking corpse', syndrome, in which the person has the delusion that they are dead or putrefying, even claiming they have lost their blood or internal organs. Both Cotard's syndrome and post-natal psychosis can manifest as psychotic depression, and I couldn't help wondering if they were connected here, with morbid thoughts of decomposition prompted by Lisa's extreme experience of seeing dismembered bodies. 'Did she mention the helicopter crash to anyone when she booked into the antenatal clinic?' I asked my colleague. She had not. It had taken another life-changing event – albeit something positive – to bring this acute response to the fore.

I was sorry and fascinated to hear this. As it happened, I was embarking on some research into maternal violence and its motives, and discovering more about how 'matrescence' – the transition to motherhood – can be a potent psychological stressor, even when the pregnancy and the baby are much wanted. This isn't widely appreciated in our culture, partly because of the societal convention that women are meant to prosper and 'glow' if they are lucky enough to conceive.

My continued research in this area and my work in the family court would reinforce the idea that there is nothing like a pregnancy for activating memories of being vulnerable and making mothers-to-be aware of the potential fragility of life. Rather than 'grounding' women (the earth mother ideal), maternity shifts self-perspectives and memories, like tectonic plates moving in the deep unconscious. This is disorientating for both woman and baby. As one struggling young mother described it so well, 'I'd been so busy looking after my body I forgot to look after my mind.'[9]

I told the psychiatrist that there had been no sign of any persistent mental health problems when I interviewed Lisa. She had successfully avoided engaging with her sense of fear and helplessness back then, which meant I saw only her briskness and lucidity, her faith in a positive future and her 'just the facts please' approach.

With hindsight, when I recalled how she described her sense of helplessness and that distorted sense of time in the absolute silence after the crash, I realised I had missed something darker, possibly related to what she'd seen of Terri's death. Years later, the experience of her body opening up during labour and her aware-ness of her new baby's helplessness and vulnerability might have activated an embodied experience of deep terror. Lisa's predica-ment stunned me, leaving an imprint on my memory to return to in future. How remarkable the process of encoding memories and retrieving them can be; the more I learn about this, the less I think I know.

My experience with Lisa also made me ask myself whether I had any feminine or cultural bias that had encouraged me to want to see her as resilient, as another young woman working in a largely male world. She had described herself to me as 'quite the little mother'; why did I (and our culture in general) think that was such a good thing? There's a serious flaw in the notion that those who nurture others are automatically strong themselves. If that were true, not only mothers but doctors, teachers, child carers, leaders and mentors would be unshakeable – and we know that's not the case.

The perinatal psychiatrist and I kept in touch, and in time she told me that after specialist treatment and care, Lisa made a good recovery and went home with her baby girl. Lisa's story made me reconsider some important ideas. For starters, I would be more thoughtful about taking at face value how people speak and pres-ent themselves. I needed to bear in mind that not everyone who speaks briskly and chattily about the past will always be resilient. I'd had a theoretical knowledge of this when I first read about the crash, but meeting Passengers A, B and C added so much emo-tional colour and nuance to my understanding.

Lisa's story stays with me even now because of the mystery of that acute contrast between her response to the crash and her reaction to becoming a mother. There is good data indicating that a minority of people do not suffer from persistent memories or intrusive thoughts after a traumatic event, so was she just lucky with helicopter crashes and unlucky with the transition to a new identity as a mother? Or was there some definite link there, a kind of delayed reaction to the physical and emotional life changes she was experiencing, not unlike my POW patient, Tom? I never saw her again, so I can't know the answer. I am left to wonder whether traumatic memories somehow hide in the mind. But if so, where? All I can say with certainty is that recovery from trauma sets its own timeline, a trail that may bend, incline or twist without warning.

The Last of the Line

Middle age is a time when we are prone to depression. At this stage in life, many of us, male or female, begin to be aware of time's swift passage, and arriving in Dante's 'dark wood', we may feel a little lost. But most middle-aged Welshmen with depression are not referred to a specialist trauma clinic, and Evan H was also one of the very few patients I'd seen for whom the trauma of the Holocaust was an active issue.

My professional training as a psychotherapist was nearly complete, and I would soon leave the Middlesex clinic to work full-time as a forensic psychotherapist at Broadmoor. Given that my time in the clinic was limited, I needed to be thoughtful about which cases I took on for treatment, but I felt compelled to hear more about this one. This man had only recently found out that his parents had narrowly escaped from Germany at the start of the Second World War and that their families had later been arrested by the Nazis and murdered in the concentration camps. He was struggling with depression and had asked for a referral to our clinic. The referral explained that Mr H was moving to London soon, making it easier for him to pursue treatment. The GP added a note that his patient, who was married with two children, worked in the field of medical genetics, studying inherited fatal diseases. This immediately intrigued me, given that his problem seemed related to a tragic inheritance from his parents.

I had a deep interest in the Holocaust that was not solely academic. Like so many, I had been moved by seminal accounts from people like Viktor Frankl and Primo Levi, and as a forensic professional, I was struck that they had been treated and labelled as 'offenders' against the Reich, purely on the basis of their identity. In my psychotherapy training, I was constantly coming across writers with lived experience of Nazi persecution: Sigmund and Anna Freud, forced to flee Vienna; Bruno Bettelheim, who survived Dachau; and Sigmund Fuchs, who escaped from Vienna to England, changed his name to Michael Foulkes and became the father of group analysis.

I also knew people who had survived the Holocaust. My father's best friend had escaped with his family from Germany, and the father of another friend arrived in England on the *Kindertransport* (the British operation in 1938–9 that enabled thousands of children between the ages of five and seventeen to escape from Germany after Hitler came to power). Some of my closest friends had stories to tell about parents who'd had their childhoods disrupted by forced migration from the terror of Nazi invasion. I also learned a great deal from a fellow trainee psychotherapist, Alfred Garwood,[1] who had actually been born in Belsen.

I wanted to know much more about how people survived terrible events and at what cost. It was plain to me by this time that there was no one-size-fits-all trajectory out of horror into healing. There are many elements that might influence outcomes, from someone's early childhood attachment security, as I'd seen with the helicopter crash survivors, to isolation from home and community, as with Nadia and Besim's story. Survivors of the Holocaust, like refugees or soldiers, are not a homogenous group, and whatever they had endured, the quality of their survival could sometimes so surpass expectations that it took my breath away – especially if they'd been children at the time. I remember hearing about the life of Erik Kandel, whose parents managed to flee Vienna after the *Anschluss*, leaving nine-year-old

Erik and his brother behind to come separately later. These children endured severe threats, early childhood separation and the loss of everything they knew and held dear, compounded by a frightening journey under harsh circumstances to join their parents in America. It would be easy to assume the worst possible outcome, yet Kandel went on to win the Nobel Prize for his pioneering work on learning and memory.[2]

So, what of the children of such children, this next generation who were now adults in the late 1990s, many of them approaching middle age? The idea that parents will transfer their pain to their children is not new; children model themselves on their parents for good and ill. We're not made of marble; we're semi-permeable, complex organisms, more like bindweed, entwining with those around us as we grow, flourishing in the sunlight and struggling to thrive if there are long shadows – the horticultural metaphors abound. The idea that the Holocaust might cast such a 'long shadow' on family life was first described by Canadian psychiatrist Vivian Rakoff, who found high rates of what he called 'intergenerational trauma' in a small sample of children of Holocaust survivors. He made the startling observation that 'it would almost be easier to believe that they, rather than their parents, had suffered the corrupting, searing hell'.[3]

By the time I met Evan, research interest in transgenerational trauma was on the rise. I'd attended lectures by Rachel Yehuda, who led pioneering research into how chronic exposure to traumatic stress might cause genes to change their shape. She and other researchers would explore how high levels of stress hormones in pregnant women affected by the Holocaust could affect the genetic make-up of their offspring, making them more vulnerable to PTSD after being exposed to further stress.[4] It was also clear that the mechanism was highly complex and unlikely to be the same for everyone. I wondered what story this man would have to tell.

———

He was tall and as broadly built as a rugby forward, but as I watched him climb the stairs to my consulting room at the clinic and take his seat, I had the impression that every movement he made was an effort. His expression was anxious, giving him a slightly wild look, although he was neatly dressed and punctual. He thanked me for seeing him, and then, without further preamble, announced he'd been feeling 'rubbish' ever since his last remaining parent, his mother Ruth, had died the previous year. At first, he'd assumed his response was typical for anyone in mourning, but after he found some letters and documents that suggested there was a painful backstory in his family that he'd never known about, he started to think differently. Now, he was applying his research skills to try and understand what was happening to him, recognising that there must be some link between his parents' trauma and his mental distress. I heard a bitterness in his voice when he added that he couldn't believe he was learning about their story only now, halfway through his life. It seemed they'd barely shared anything of their past with him when they were alive.

I told Evan something of what I knew about intergenerational trauma. Much seemed to depend on how survivor parents coped with their natural grief, rage and loss of identity, and what they communicated to their offspring about their experiences. I gave as an example Anne Karpf's *The War After*, her memoir of living with parents who were Holocaust survivors and who did tell her about their past, and the challenges she faced as a result.[5] Evan then shared with me the work of Yael Danieli, a clinical psychologist and director of the Group Project for Holocaust Survivors and Their Children in New York. It felt almost like we were speaking as colleagues rather than patient and doctor, and I wondered if this was deliberate on Evan's part, a way of delaying getting to the personal. Given this was our first meeting, that was not unusual, but maybe he also felt more comfortable in lecture mode, intellectualising his distress.

'Danieli's view is that all children of Holocaust survivors will be affected by their parents' experience,' Evan said, adding, 'Sort of like a lit fuse of unknown length – it's just a question of when and how the bomb explodes. Which makes total sense to me since I'm working on some research into Huntington's . . .' He paused to make sure I was with him. I nodded to let him know I understood about how this life-shattering illness is passed across generations by a dominant gene. I told him I grasped the irony of him dealing daily with fatal physical diseases that can be silently passed from parent to child without anyone knowing. I couldn't help but muse aloud about whether his choice of study had been influenced by an unconscious awareness of the dark family history that had been withheld from him. He looked surprised when I said this, but nodded, saying, 'That never occurred to me . . . I'll go away and think about it, Gwen. Sorry – can I call you Gwen?' I told him that was fine; there usually comes a time in therapy when patients may ask to use my first name, although it doesn't happen often in the first meeting. Again, it was as if he preferred to talk to me as a colleague, protecting himself from the vulnerability of being a patient.

We agreed to meet again and talk further. Evan was clearing out the family home as he prepared to move to London, and he would look for more of his parents' papers. His step seemed a little lighter as he departed, but I was left with a worry that he and

Danieli might be right: that his mental difficulties went beyond mourning and could plunge him deeper into grief and loss.

———

It was our third or fourth meeting, and this time, Evan had brought with him a large, battered old leather briefcase. 'This was in the loft.' He rummaged in the bag, giving me a view of thick dark hair that was beginning to streak with grey. Then he pulled out a folder and extracted a piece of notepaper. It was a crooked photocopy of what looked like an index card, which someone had annotated, in a cramped hand, 'Internment Tribunal Card: Heller'. He glanced up at me. 'My parents changed their name from Heller soon after they arrived.' I made a note, knowing that was not uncommon at a time when a Jewish-sounding surname could be fatal in Germany and German-sounding ones were met with suspicion in England. He handed me the page, and I scanned it briefly. David and Ruth, 'both of Jewish race', both studying medicine. In the summer of 1939, when they were newlyweds, the Nazis detained their parents and siblings. The couple immediately left Berlin and crossed Europe. By the end of December, they were registered at an internment centre on the Isle of Man. The few lines of bureaucratic prose on the card masked a world of pain.

Evan reached down again and pulled out a couple of fat folders. 'We found more letters, all sorts of other documents, too . . . Look at this. She must have kept everything from when they got to Britain. But all these years, I never knew any of it was up there, never set eyes on it . . .' He broke off. 'Sorry, can I have some water? Dry throat.' I poured him a glass from the jug on the table between us and handed it to him. He gulped it down, eyes squeezed shut. I wasn't sure how far he wanted to go today, but we had time. I felt more like his assistant cartographer than his therapist, keeping him company in the field as he began to carve his way through unknown rugged terrain.

I gave him the card back, and he studied it momentarily. 'It's the details they gloss over that get me, like this bit: "Made their way from Berlin to Paris and on to Calais on foot." They probably wouldn't have risked taking the train. So what did they do? Walk the whole way? In a matter of weeks? Can you imagine? With war breaking out and winter coming . . .' He sat back and pushed up his glasses to rub his eyes. 'It makes me tired just to think about it. I mean, what did they eat? Where did they sleep? Did they dare stop anywhere to rest?' There was an edge of panic in his voice.

'It must have been terrifying,' I agreed. Our conversations often went like this; together, we would feel our way through the incomprehensible as best we could. It had been clear to me from the start that in Evan's case, neither a simple diagnostic label of depression or PTSD nor conventional treatment would be helpful. Rather, he needed to talk to someone who could listen for what he might not say. Despite his evident distress, I thought it was good that he was asking questions about his parents' past: curiosity is a sign of people opening their minds and allowing in new thoughts that might change their perspective.

Evan wanted to tell me about a recently established Holocaust archive that he'd been able to visit in London, which was gathering video and audio testimony from survivors. 'I came across an account by a woman who'd made pretty much the same journey with her husband. She described how she looked in a mirror when they finally got to Calais and didn't recognise herself. She'd left Paris weeks earlier as a healthy young woman with dark hair, and now all she saw was a ghost, thin, grey and pale.'

I remembered that when we first met, he told me that after his mother's death he'd lost a lot of weight and 'kind of aged over-night'. This did not seem unusual for someone of his age who was experiencing a parental bereavement, but today, his gaunt features and grey affect struck me differently. A grey ghost. What a poignant idea: trauma can vaporise us. Paradoxically, our grief for lost loved ones makes us ghosts. Did Evan think his mother seemed ghostly – or haunted – as she went through her life?

'I don't know. Too late to ask her now.' Again, there was some anger in his voice, although he quickly tempered it with compassion. 'God help her, she was only twenty-three years old! On the run, surrounded by danger and fighting, her life destroyed before she'd had a chance to live it.'

'Have you found out what happened next?'

Evan opened a folder, sifted through a few papers, and then held up a letter. 'They got a sponsor.' He read from the embossed letterhead: 'Dr Robin Evan D, General Practitioner, Monmouth.' This man had testified to their character and valuable skills as able young medical students who could help with the war effort. Evan read out the final line in a florid accent: 'I wholeheartedly assure the esteemed committee I have no doubt as to the Hellers' absolute loyalty to our king and country. I live on my own and shall be pleased to welcome them to my humble but ample enough home in Wales.' He glanced up at me with a brief smile. 'Never used one adjective when three would do. He was quite a character. A good man, Uncle Robin.'

I was confused about the connection. 'How did your parents know him?'

'He was a friend of my grandfather's – my mother's father. This is one of the few things that my mother did tell me. She said they met as students, when her father came over here as a boy to improve his English. They taught each other to swear in their native languages.' Evan smiled slightly. 'The mind boggles. Welsh and German curses. Think of all those consonants.' He paused. 'As a kid, I always thought he was my real uncle. Or great-uncle. He was pretty old.' He snorted. 'Probably the age I am now, come to think of it. Anyway, even when I found out we weren't related, I still called him Uncle Robin. It made me feel more . . . normal, I suppose. We had no other relatives. He died when I was in secondary school. I was named for him – my parents wanted me to have a Welsh name.'

I let him talk and talk; he needed to. He was grappling with an utterly changed perception of the relationship between himself and his parents and the awkwardness of a new sense of self as the son of Jewish Holocaust survivors. In a small way, I thought I could identify with this process, as I was sometimes finding my new identity as a psychotherapist unsettling – although, clearly, the contexts were very different and Evan's new perspective on his identity was far more disturbing and disorientating.

Early on, he talked to me about how hard he was finding the move from Wales down to London. It was a fresh start, but he had a strong attachment to the place where he'd grown up with loving parents. Despite their unwillingness to discuss the past, he had always felt seen and cared for by both of them. When Uncle Robin passed away, they inherited his house, which sounded like a place of safety and security for all of them. After university, Evan returned to live with his mother and father, and when he married his wife Emma and had children, he and his family settled nearby.

His first experience of bereavement was when his father died of cancer some ten years previously. He supported his mother in what he thought was a natural process of grieving. Then her mind began to collapse, and the radical change in her took him aback. It was moving to hear him describe the disappearance of the mother he had always known and the emergence of strange fragments of a woman he did not recognise. She began speaking to him in German and sometimes cried for people whom he knew nothing about. After she died and Evan's wife was offered work in London, they decided to leave Wales. He welcomed the change but felt the rupture of this final connection with his parents.

He was eloquent about his feeling that there were two aspects of himself coming to an end: a Welsh self, which had seemed so benign, familiar and ordinary; and a parallel Jewish self, which had lived 'in the dark' (his words). I picked up on that, recalling the Theodore Roethke line: 'In a dark time, the eye begins to see.'[6]

'What does being in the dark mean to you, Evan? Is it frightening?' He thought for a moment. 'Sometimes stepping into the light is worse – it's blinding.' He felt constantly exposed and vulnerable lately, he said, and the anxiety this caused was all the greater because he could not turn to the two people in his life who had always provided him with safety and comfort. Their long deception (as he saw it) was painful and confusing, and it was difficult to reconcile these antagonistic feelings with the grief. So that is what we tried to do, session after session, in my little room upstairs at the clinic, meeting each Monday in the early evening. Working with him brought home to me how essential it is for a therapist to be willing simply to bear witness to whatever people needed to 'bring to light', however intimate or painful it might be.

Over time, he began to read aloud his parents' documents and letters, an experience which was sometimes unbearably poignant. They were all written in German, which, Evan told me, was somehow surprising to him. He could barely decipher a word of the spidery script in the originals, but a friendly colleague in the language faculty at the London university where he now worked had agreed to translate a selection of the correspondence.

One day, Evan brought with him a packet of love letters between David and Ruth. 'How do you feel about reading these?' I asked. I didn't know what this might be like for him, and I glanced at the clock, wanting to make sure we had enough time to reflect on any thoughts and feelings that might arise. Much later in my life, when my own parents were gone, I came to realise how complicated it is to hear your parents' younger voices in their correspondence and the awkward, intrusive sense of peeking into their private realm. I doubt I'm alone in feeling that such letters can activate old attachments, bringing the past into the present with an intensity that is not always comfortable.

But he was keen to proceed. Most of the translated letters dated from the early 1940s, when the young couple were separated for long periods. Ruth stayed in Wales, assisting Uncle Robin in his surgery, while David was volunteering in a London hospital and trying to complete his training to qualify as a doctor. Theirs seemed to be a relationship of deep affection and care, interwoven with constant anxiety about the bombing raids on London and whether they would meet again. David wrote about spending many boring hours huddled in Tube stations, sheltering with his colleagues during the frequent air raids on the capital. In one memorable letter, he described how he had discovered a new way to keep his fear at bay during the bombings: he would relive scenes from their life in Berlin, returning to the time when he and Ruth first fell in love.

The irony that David could cherish these memories of Germany even as his countrymen continued to try and kill him seemed absent from the letters. His courage reminded me of Viktor Frankl's *Man's Search for Meaning*,[7] a book I thought Evan might wish to read when he was ready. At one point, Frankl describes the pure agony of Nazi soldiers marching Jewish prisoners through the snow, beating anyone who moved too slowly. A man beside him had whispered, 'If only our wives could see us now! I do hope they are better off in their camps and don't know what is happening to us.' Immediately, Frankl was able to picture his wife, her smile and her voice in his mind offering him blessed distraction and strength. His love for her helped him to survive three years in a concentration camp and to deal with his grief when he realised that they would never meet again.

Evan read aloud his father's account of how he and Ruth had first met. The letter recreated the memory of a fateful morning, when David had come down for breakfast in his family's home, a large townhouse in Berlin. He heard someone playing the piano in the morning room, and throwing open the double doors, he startled Ruth, then just eighteen, who was waiting to walk to school with Leah, David's sister. 'She jumped up,' said Evan, 'and according to Dad, when their eyes met, he had this odd sensation, what he calls "a shock of recognition", although they'd never met before.'

'Like a thunderclap, love at first sight?' I smiled.

Evan smiled back. 'It sounds corny, but to me, knowing him the way I do, how passionate he is . . . I believe him.' I noted the use of the present tense and could see that his emotions were close to the surface. After a moment, he gathered himself and read another letter aloud. In this one, it sounded like David had been trying to encourage his young wife to stay strong, writing something along the lines of 'God hasn't brought us this far only to drop us', which prompted me to ask about their faith. Evan shook his head. As a child growing up, he only remembered celebrating Christmas and Easter as a family, just like everyone else in their small community. As far as he knew, they'd never set foot in a synagogue nor marked the Jewish holidays.

'What are your mother's letters like?'

'She didn't write as much as he did. I think she was always more of a visual person than a writer.' His face turned serious and sad. 'There was one, though. I'll bring it along next week.' It sounded important: in this letter, she spoke about her longing to start a family as soon as the war ended. She hoped for two girls and suggested to David that they could name them for the sisters: David's sister Leah and a younger sister of hers called Hannah – another piece of the jigsaw puzzle of their past, falling into place. 'Growing up, I never heard her mention these names, these sisters, ever,' said Evan. In his recent research, he had managed to determine that both girls had been arrested with their parents and transported to death camps, where they were murdered. 'I always believed my parents were both only children, like me. I'm not sure they said so explicitly, but sometimes a lack of information creates its own story, doesn't it?' I could only agree and had to stop myself from speculating about how it was that David and Ruth got away while their sisters were arrested, and to what extent they suffered from survivor's guilt, a syndrome first – and specifically – documented in the 1980s in survivors of the Holocaust.[8]

'Anyway, Mum never got her wish. Only me.' I thought that sounded important, as though he felt keenly his mother's disappointment that he was not the child she had wanted. What had stopped his parents from having more than one child? Was that a choice? Evan admitted these were among the questions he'd never asked. He hung his head, and his sorrow darkened the room.

I sensed he was in such pain about the unknown and the unasked that it might be best to shift our dialogue to the present for now. Could he tell me about his children? He brightened immediately. 'Emily and Lucy. They're lovely . . . I mean, they were a bit stroppy initially about the move to London, but now they're adjusting, and we've promised to have lots of holidays in Wales so they can keep up with their mates, and I'll want to go back to visit my parents' graves . . .' I couldn't help but notice how quickly he'd returned to the theme of loss and death.

'What ages are the girls?' I asked. I wanted to see if he could stay with the living, but he barely lasted a sentence. 'Twelve and fourteen. I adore them, but sometimes I wish we'd had a son, too. When Mum died, I realised I was the only remaining connection to the Heller family. When I'm dead, that's the end – I'm the last of the line. I hold so little of the family, just these pages . . .' He indicated the letter in his hand, then set it aside. He looked miserable. 'I feel so adrift, Gwen. I have so many questions about my parents' past lives, and the people and places they refer to in their letters all seem unreal to me, like something from a novel or a film. I don't understand why they didn't tell me anything, nothing at all. Ever. Okay, I get why they might not have when I was little, but why not later, when I grew up?'

I couldn't know the answer to that, but I immediately thought of the work of the American theologian and author Frederick Buechner, whose thinking has long inspired me and been a source of insight. In *Telling Secrets*, his 1991 memoir, he explores his father's suicide when he was a child, a close-held secret for years, proposing that 'the human family all has the same secrets, which are both very telling and very important to tell'.[9] Buechner was also involved with Alcoholics Anonymous, where participants are encouraged to let go of their resentments, on the basis of the old adage that 'we are only as sick as our secrets'. I did not see Evan as someone with a medical diagnosis, but it sounded as though his family secrets were making his heart sore and activating a deep longing for his parents to be there to make things better.

I asked what story of their lives he *had* been told as he grew up. 'We were British,' he said flatly. 'Yes, they had slight accents, but their English was excellent. They were proud to be British. They said they'd come from Berlin before the war, looking for job opportunities, after their parents died. In a dreadful way, that was true . . . So I guess they didn't lie to me, they just didn't tell me everything.' Or couldn't, I thought to myself. I tried to imagine how these two young people might have felt about putting their experiences into words for their son. As parents, did they want to protect him from their memories of fear as they escaped to safety, the shame of internment and the eventual horror of learning what had happened to their parents and siblings? Like Nadia with Besim, how could they say their loved ones' names or tell their tragic stories?

They had clearly decided to fully inhabit their British identity, and Evan said he never recalled hearing them speak any German in the home. 'I suppose they might have done, in private, but honestly, I never heard either of them do so around me.'

In my forensic work around this time, I was setting up a therapy group for homicide perpetrators and had come across the work of the poet Paul Celan, a Holocaust survivor whose parents had been killed in the camps. He had been reluctant to speak any German afterwards, saying the language of his mother had become 'the language of murderers'.[10] Could that explain why David and Ruth Heller had not spoken German in their adopted country? I understood their son's anger and frustration, but I also felt for them and the impossibility of helping a beloved child understand something so vast and dark. Would it have been 'better' if they had? Who could judge them, and according to what parameters?

'And when she finally did speak some German? What was that like?'

'Dad was in and out of hospital at the end,' he said. 'Mum would lay her head on the pillow next to his and whisper to him for hours. Eventually, I realised she wasn't speaking in English . . . and I couldn't understand anything. Except the love.' He cleared his throat, his voice thick. 'Then, when he was gone, she started to go downhill and . . .' He couldn't finish. No longer the lecturer, sorrow was derailing his language, as it had done to his parents before him. I felt I was watching their intergenerational grief play out in front of me.

After a moment, I offered gently, 'And then?' He gave a deep, long sigh. 'Soon after we moved her to the care home, Mum started seeing things, hearing things that weren't there . . . It was awful.' Some days, she wept for prolonged periods or began flailing and flinching from tormenters he could not see. 'She had Lewy body dementia, and visual hallucinations are common, you know? She was so distressed, I didn't feel I could ask her what was happening in her mind.' He shook his head. 'Then Emma and I began considering whether there was more to it. It seemed like actual bits of memory were rising, overflowing their container. Emma's a memory researcher, did I ever tell you that?'

He hadn't. The coupling of their professions was notable; the memory researcher and the geneticist made a potent team for confronting a loved one who is suffering from dementia (from the Latin *demens*, 'moving away from or out of their mind'). But Evan was looking past me, stuck in his remembrance. 'It's so disorientating, so sad . . . Mum's there, sitting up in bed, eyes wide, referring to our Lucy as "Leah", grabbing at her hands, trying to kiss her, scaring her, I think.' As he described this, I noticed he seemed panicky and a little short of breath. 'Scaring you, too?' He met my gaze. 'Yes. I felt . . . as if I ought to know what was happening. I couldn't reassure my mother or my daughter, and that felt horrible.'

126

———

American author, lecturer and palliative care expert David Kessler,[11] who worked with Elisabeth Kübler-Ross on her theory of the 'five stages of loss', writes that 'Each person's grief is as unique as their fingerprint, but what everyone has in common is that no matter how they grieve, they share a need for their grief to be witnessed.' Perhaps Ruth needed that from her son and his family at the end of her life. I would continue to listen to these painful stories week after week, supporting Evan in his expressions of loss, sadness and anger. Although good data was emerging about the value of therapy for trauma survivors, including those with transgenerational trauma, sometimes I came away from our sessions feeling helpless. In the face of the enormity of the Holocaust and its generational impact, how was it possible for any kind of therapy to truly change such deep-rooted distress? I would learn that this feeling was an essential part of the work – that sometimes a therapist needs to act as a container for their patient's unbearable emotion. I was holding some of Evan's sense of helplessness.

In one memorable session, he came in with a large scroll tied with string. Without preamble, he unrolled it onto my desk and carefully set a mug and a paperweight at each end to stop it from curling back in on itself. It was two sheets of card joined together with yellowing Sellotape. At the top were the carefully lettered words 'MY FAMILY BY EVAN H, YEAR FIVE'. The image below was somewhat faded, but I could easily make out that it was a tree, rendered in a child's hand. A spreading oak perhaps, its fat trunk sprouting several irregular branches right and left that were decorated with an array of large oblong leaves cut from contrasting coloured paper. The name 'Evan' was on the top leaf, and either side were leaves inscribed with 'Mum – Ruth' and 'Father – David', and nearby, 'Uncle Robin'. On the branches below were more leaves, adorned with some Welsh names I can't recall – maybe a Dafydd, an Angharad or a Gwilym. Looking further down the tree, I saw some other names that are much easier for me to remember all these years later: Lucy, Susan, Peter. Then, as my finger traced the trunk to the bottom, it came to rest on a low-hanging, precariously attached leaf that read 'Caspian'. As a lifelong reader of C. S. Lewis, I could not help but ask with a smile, 'No Edmund? No Aslan?'

Evan grinned. 'Hah. I was a huge Narnia fan as a boy. I still am now. Bit obvious, right? It was all invention. When we were given this assignment at school, I instinctively knew I couldn't ask my parents about our relatives' names. I didn't even try. I just borrowed from my friends and their families, then added some of my favourite characters for good measure. I don't think my teacher batted an eye. Look.' He pointed to the bottom-left corner, where someone had written in careful cursive, 'Well done! A+.' He paused for a moment. 'I'm just thinking . . . maybe Uncle Robin told the school something? It was such a close community. It could be there was an unspoken agreement not to ask too many questions of people like us. A conspiracy of kindness, you might say.'

I gazed down at the image of a little boy's fantasy family blossoming under his hand and thought of the unnamed – the extended family who would have loved to have known Evan, who would have loved to have lived but were destroyed, erased. I felt a wave of grief on hearing that he had known he couldn't ask his parents about them, that he didn't dare to penetrate their wall of silence. The same grief may have come up for Evan: in a swift movement, he swept the sheet up and away, rolling it and stuffing it into his bag. 'Mum must have got it from the school; it was tucked away with all my old reports. Something else she never mentioned to me.'

I looked at him, hoping he could stay with this piece of his history. '"A conspiracy of kindness," you said. How does that conspiracy seem to you now?' He frowned. 'I don't know. All I know is I always wanted to be like everyone else, like my friends. They had these big families. My friend Dai was one of six; loads of people milling about, lots of noise . . . and it was so different at my house. Quiet. The three of us would talk a little over meals and maybe watch TV, once we had one. Or Mum might play the piano. Sometimes Dad would put on one of his jazz records. He loved jazz.'

'What type of jazz?'

'All kinds – John Coltrane, Thelonious Monk, free-flowing stuff – but he also liked swing, big bands, and he loved vocalists, American greats like Ella Fitzgerald, Sinatra . . . He had a lovely voice, too. I have a memory of him singing "Fly Me to the Moon" to my mother in the kitchen, and her laughing and swatting him with a cloth, telling him not to be so foolish . . .'

'He sounds light-hearted. Was he?'

Evan considered that. 'Mostly. One of the only times I ever saw him angry or upset was to do with music. One day, we were out together, just the two of us, and we passed a big record shop. I was in my teens and begged to go in and browse for a while, but then they put on classical music, playing it really loud. Sounded like marching music, I thought. Grand. Impressive. But Dad froze, with this weird look on his face – I'd never seen him like that. He grabbed my hand and pulled me out of there, fast. I was astonished; it was so out of character. He just walked faster and faster, putting distance between himself and the shop, and I had to run to keep up with him. After a few minutes, he calmed down and apologised, explaining that he just hated that piece of music. "It's not good," he told me. "Jazz is better."'

I was fascinated. 'What was the music? Do you know?'

'It took me years to find out, and again, it was by chance. I shared my parents' love of music, and although I don't play an instrument, I've always sung, since I was at school. I joined our local choral society – one of those big male Welsh choirs, you know? One time, we're practising some pieces for a concert, when Bill, the choirmaster, plays us a tune on the piano, and I recognise it instantly: it's that one from the shop.' Again, his move into the present tense suggested that this memory had a particularly strong emotional charge. 'This sounds a bit mad,' Evan said nervously, 'but I must admit, I immediately tensed up and looked around for Dad to make sure he couldn't hear it. He'd been dead for a year. I asked Bill what the piece was, and he says, "You don't know it? It's from Beethoven's Ninth – the 'Ode to Joy'." We started to learn it in choir, and I loved it. It was so beautiful. Yet Dad had hated it so much.'

Immediately, I could hear the melody in my head, a thing of such beauty and feeling – all the more so because, miraculously, it was composed by a man who had gone completely deaf.[12] It had such an uplifting effect on me, like it had on Evan. There was a connection to Germany, but also to peace and hope. I vividly recalled hearing the famous Christmas concert broadcast from Berlin, where a multinational orchestra led by Leonard Bernstein played the Ninth after the Wall came down.[13] But I also knew that Hitler and other repressive regimes had adopted it as a nationalist anthem; was that what had set Evan's father off?

Of course, it shouldn't have surprised me. Since time immemorial, music has been a way for people to capture and encode emotional communications that have great meaning. It may have pre-dated speaking as a form of communication because, as social animals, we need to be able to share emotions, and music can communicate what we are feeling and build social bonds. Like laughter, there is no culture or country that does not have music, dating back to the earliest humming sounds thought to have been used by the Neanderthals.[14] Music's job is to convey and help us manage our feelings; in one recent American study, researchers mapped as many as thirteen key emotions that are triggered when we listen to musical performances, no matter the genre.[15] We all keep lists in our heads of our favourite (and least favourite) pieces of music, and each and every one is a potent activator of memories, good and bad.

Evan was lost in thought. 'I suspect Dad hated anything that recalled the High German classical values he'd grown up with and how they were lost. I'm guessing *that* music was a reminder of the destruction of everything he loved. What I remember of him was his total interest in the future, in science and medicine, and in modern music. These were the things that gave him hope and joy, and they were part of what he wanted for me, too.

'At the time, I figured he just didn't want to talk about it. But later, surely he could have tried . . . They could have talked to me about that – and everything else, too. I was an adult, for God's sake. Or maybe by then, they thought it was too late.' He shook his head. 'Gwen, since I've been coming here to talk to you, I've realised that I always felt there was some unspoken barrier between me and my parents. I keep asking myself why I never instigated a conversation about it. It's like I picked up on some warning signal that it might be risky to do so, you know?' He added that in his research, he saw how this happened sometimes in Huntington families, when someone died, and people didn't talk about why.

'Not just out of grief, but because the legacy is so terrifying.' I was fascinated by him drawing a parallel between that disease and the Holocaust – there was something important to contemplate later. But for now, I wanted to follow his emotions, his sense of regret. Could he say any more about what he was feeling?

'It's all jumbled up,' he said, feeling his way as he tried to answer. 'Initially, when we found all the letters and stuff, I felt sad and angry and bewildered, but now . . . now I think I have a better understanding. They had a right to their secrets. They loved me. I never doubted that. They probably wanted me to have the security and love they seemed to have known as children, before their world fell apart. So they buried the awful twist their lives took as young adults—' He stopped there, but I sensed he wanted to go a little further, and I waited for him to continue.

'We all lived with something sad and unsaid in that house – something . . . unclear.' Unclear. This was interesting. I reflected that he'd embarked on this journey of discovery while 'clearing' the house after his mother died, when he found the cache of papers in the loft, including his parents' letters. I thought he'd also gained clarity by bringing concealed things into the light of our meeting room. He agreed, adding that he felt sure he wasn't finished yet and needed to dig deeper still to know more.

———

The next time we met, it was our last session before the summer break. He was full of news, telling me he had unearthed more about the family history in Berlin. Apparently, four generations of the Hellers had lived in the same house, right up until his family were arrested and deported. The Nazis had confiscated it, and after the war, it had become a hotel. 'Really fancy, apparently.'

Evan's tone was brusque. 'Nothing like the house I grew up in.' The Heller men had all been doctors, and like Evan himself, there was an uncle who had been a medical researcher and teacher. All accepted, valued members of society, dedicated to serving their communities, never dreaming that those communities would first exclude them, then attack them, and ultimately murder them.

Evan had been busy. He'd also visited the German consulate in London, where he found out he was eligible for German citizenship. This sounded significant. 'Do you want to pursue that?' He wasn't sure. If he did embrace this new identity, which offered a closer link to his parents and their families, it would be a big step. I asked if he was considering going to Germany – he'd told me he had never been before. Another step towards clarity?

'We're talking about it. We might do that in August. I'd like to see their house, even if it's just to look inside, you know? I don't think we want to pursue a reparations claim, though. We talked to a lawyer, and it sounds awfully complicated: because so many records were lost in the East, it's harder to chase up the chain of title and all that. I think more than anything, I want my family, Emma and the girls, to see. Just to see.'

'What do they think about that idea?'

'Until recently, the girls thought this was Dad's stodgy genealogy project or something. But now they've begun to ask more questions about the Holocaust. They're curious. And Emma's all for it.'

I commented that I'd been to Berlin several times and had been struck by how simultaneously close and distant the city's dark history felt when I was there. 'That's what I'm hoping to see,' Evan said. 'It's modern, there's been lots of rebuilding, but they say the past is available to explore. I honestly can't believe I've never thought to do this before.'

'Perhaps it wasn't possible while your parents were alive? Maybe that felt like a border you couldn't cross?'

'More like an electric fence.'

I was startled by that simile, so resonant of concentration camps to me, but I thought I understood. Talking to his parents about their past was something *verboten*; at some level, it had felt risky, or even fatal, to venture in that direction. Evan sighed, slumping in his chair, looking more like his old, grey, ghostly self. He covered his face with his hands. 'Oh, I don't know, maybe . . . maybe going over there is a bad idea. I wish . . .' His voice was muffled.

'You wish . . . ?'

He lifted his head, looking at me. 'I'd give anything to tell them, to *show* them how much their suffering means to me. I could have helped them bear it, maybe. And I wish I could tell them about this discovery process I've been on, my reconnection with the Hellers and Germany. Would they want me to try and make a claim on my father's house? I don't know. I realise I sound like a broken record, but I *still* just wish we had talked about this when we had the chance.' His eyes were wet, and I could see the deep anguish there.

At times like this, the words therapists use often feel inadequate: 'I see'; 'I hear you'; 'That sounds hard.' But what came to me instead was, 'How much things have changed since our first meeting, don't you think, Evan? You were in transition, what with the move from Wales to London and losing your mother. Over time, I've seen that you've been able to do something your parents never could: articulate difficult feelings, make connections between your family's past and present, move towards a new identity that embraces being German – and perhaps you're bringing some new quality to being "the last of the line"?'

'How do you mean?' He frowned. I was thinking, but did not say, that he might be wearing that label like a hair shirt at this point, and he might even be reluctant to give up the familiar pain of it. I tried to find the right words. 'To be last sounds like a bad thing, on the face of it. But the last person can also be the beneficiary of so much learning and healing from those who've gone before them. And being last, they are in a position to break a chain of behaviour or silence. Isn't that what you are doing by sharing this journey with your daughters, talking to them about the Holocaust and your family?'

Evan took that in. 'Right . . . I am doing something my parents couldn't do, showing my children that the past doesn't have to be hidden, like some dangerous dominant gene. I think they're old enough to understand that.' He paused, then suddenly exclaimed, 'I'm not leaving my children in the dark like they did to me. I love and miss them so much, but they never gave me a chance. Why, why couldn't they trust me with their pain?'

I had no good answer for him. He sat for a minute, hands twisting together in his lap, his face contorting as if he was fighting not to cry. 'I wish I could tell them I understand how hard it was . . . I understand they must have felt so alone, being young parents, with no family around to help, and everyone they loved far away, vanished.' His voice rose. 'God! It's unbearable to think about.' He pushed himself out of the chair and stood, pacing in the small space, his hands raking his hair. It was as if he didn't know whether he should scream, weep or run from the room. I wondered if his parents had also known that feeling.

So often, therapy sessions before a break will end on an acute note of sadness or anger like this. We were almost out of time and wouldn't see each other again until after the summer, after his family trip. I told him I looked forward to hearing how that worked out. I wanted to leave him with an idea I'd had, inspired by the correspondence he had shared with me. 'Evan, would you consider writing a letter to your parents about what you see in Berlin? We could think about it together here when you return, if you like?' He seemed reluctant initially, as people sometimes are about writing a letter to the dead – what's the point? But I'd known this exercise to be helpful and empowering, especially for grieving people, and he agreed to give it a try.

———

When he returned after the break, it was one of the first really cold days of autumn, and he looked much changed. His face was flushed pink as he stepped into my office and settled into his usual seat, loosening the brightly striped woollen scarf that was around his neck. No grey ghost today. They'd had a good time on their trip, he told me, before taking out the letter he'd written, holding it up for me to see, like a good student who's done his homework. 'We'll get to that,' I said, 'but why don't you tell me a little about the trip first?'

The girls had loved the city, as had his wife. 'And you?' He shook his head. 'This will sound weird, I know, but I felt like I knew my way around. I've never even been to Germany, let alone Berlin.' I said that didn't sound weird at all. If we believe parental trauma can be transmitted, might their other memories, including their sense of place, be passed on, too? We chatted a little more about this, but I could tell he was eager to get on with his letter.

'Shall I?'

'Go on,' I said.

'London, 21 September 1998. Dear Mum and Dad, we've just come back from Berlin.'

Evan stopped, clearing his throat and shooting a glance at me. 'Sorry. I feel self-conscious, reading it out.'

'You don't have to if it's too hard?' I offered.

'No, no – it's okay.'

He read on, describing in detail all they'd seen, from the Brandenburg Gate to Checkpoint Charlie, where they'd acquired little packaged slivers of the Wall as souvenirs. He broke off to tell me, with a slight smile, 'How my dad would have chuckled at that – the ultimate capitalist move, selling off the Wall, eh?' They visited art galleries and war memorials, and his elder daughter Lucy insisted on a full day at Berlin Zoo in the Tiergarten, one of the oldest and largest in the world. 'She still wants to be a vet, Mum – animal-crazy as ever,' he read from the letter, smiling. I felt his parents' presence with him as I listened; there was a beautiful sense of connection in his speech.

Despite my travels in Germany and long interest in the Holocaust, it was from Evan that I learned about the German artist Gunter Demnig, who came up with the remarkable idea of making the largest Holocaust memorial in the world. In Berlin and dozens of other European cities, he created small brass plaques inscribed with the individual names of those who had died, set at random intervals in city pavements. 'They call them *Stolpersteine*, Mum,' Evan read. He explained how the girls had noticed them first, commenting on how funny it was that kids always seem to see things their parents might avoid looking at. I thought his mother and father might have wholeheartedly agreed, if they'd been with us now. Their son was seeing the world through new eyes, after all.

Fascinated by Demnig's 'stumbling stones', the girls started keeping count as they found one, then another, pausing to read the names aloud, just as we might stop and read the commemorative blue plaques in London that remember famous men and women of the past. But the *Stolpersteine* were not celebrating the famous. 'They're these tiny monuments to Everyman – or woman,' Evan told me, with a sense of awe in his tone that I found very moving. 'It's such a simple, beautiful idea. It comes from the Talmudic principle that a person is only forgotten when his name is forgotten.' I told him I was familiar with the idea, even as I was once again reminded of my patient Nadia – a woman who had found it almost unbearable to utter her late husband's name.

Finally, on their last day, they went to find his father's house. 'It's a hotel now, in what was East Berlin,' Evan told me. In his letter to his parents, he described how when they found the place, he and his family stood outside for a time, preparing to go in. His voice faltered as he spoke about how he was trying to feel their presence and imagine them at the door. The girls, who'd been chattering away the whole time, fell silent. They all held hands, looking up at the house.

As he got to the moment when they stepped into the hotel lobby, I realised I was holding my breath, as if I were right there, too. My mind went to the letter he'd once shared with me, where David had described that day long ago when he first saw Ruth sitting at the piano in the morning room of this very house. I felt a lump in my throat as Evan went on reading from his own letter, which I remember now as something along these lines:

We were ushered into the reception area, and the manager said we were welcome to look around. She told us the place had been completely remodelled over the years, and I felt a real pang – as if I'd been cheated out of the memory I deserved. Dad, it turns out the room where you first saw Mum is now the hotel restaurant. I stopped at the threshold, imagining you as a young man, bursting in there to find a girl with long dark hair playing Liszt by the French windows.

Evan paused, glancing up at me, clearing his throat, and I made some gesture or comment to show that I remembered that moment from the love letter; we were 'on the same page'. He told me they'd roamed around the building, finding a grand curved wooden staircase leading up to the bedrooms, which they were told was part of the original building. The children were restless, and soon enough, they were racing up the stairs to see who could get to the top first. Evan's response to this ordinary moment was significant, as it released a torrent of questions I'd not heard from him before, things like: 'Dad, did you ever slide down that wide banister as a boy? Do our girls look anything like your sisters? Mum, did you get on well with your sister? How did you even manage to escape Berlin, when the rest of your family were arrested and taken away? What happened? Did someone help you? I have so many questions, big and little, things I'll never know. I don't think I would ever try to claim the house. Should I? People say I should try, but it doesn't feel like mine. Do you understand that? I wish I'd known my grandparents, as the girls knew you. Was that very painful for you, that your parents never got to know me? I feel like I miss all those people who died long before I was born, as if I love them, even if we never met – does that make sense?'

He stopped reading. There was so much emotion in the room that neither of us could say anything for a few minutes. I felt so privileged to have heard his letter. Later, I thought, I might tell him about the biographer Richard Holmes, who wrote about following in the footsteps of Robert Louis Stevenson and other literary greats, getting to know them by tracing their journeys.[16] Rather like the *Stolpersteine*, 'footstepping' caused Holmes to pause and reflect in the present on lives past (his subjects' and his own) and to think about people in relationship to specific places. And he, too, had known that rush of disappointment when a feature or a topography had changed, and he was prevented from being able to retrace every step. But when we cannot have all the answers or fully complete a connection, that allows us some space to carve a new path for ourselves into a future, away from the past.

Evan was speaking again, his voice low, almost confessional. 'As we left the house, Gwen, I had to force myself to walk. I badly wanted to run, get the hell away from there as fast as I could and put some distance between myself and that place. It felt urgent.'

This reminded me so much of his father running from the record shop. But I simply asked him, 'What do you think that was about?'

'I don't know. Panic. Too many ghosts. Just impossible to take it all in.' He shook his head rapidly to and fro, as if trying to dislodge something. 'Anyway, I managed to calm down, did some deep breathing, and we walked back to our hotel. The girls skipped ahead of Emma and me, then ran back, telling us they'd found another *Stolperstein*. Ugh, I'm not sure how to pronounce that right. I must learn some basic German. I want to know my parents in their own language.' I didn't respond, but I could see that he realised how vital that thought was. His face softened as he told me that Emily, his youngest, had proposed that they get one of those plaques and put it outside the house – 'We'd put my aunts' and grandparents' names on it. We're told anyone can sponsor them, and the city will install them. I'm looking into it.' I thought it was a beautiful idea and said so.

How had he been coping since they got back to London? Evan told me he'd stopped at a synagogue near his new house the week before and asked the rabbi if he could have a chat and maybe even attend services sometime. He was welcomed. He told the rabbi about his parents, and they prayed for them together. 'He told me that when someone dies in their community, they always say, "May their memory be a blessing." I love that so much.'

It seemed to both of us that our work together was coming to an end. I was soon to move on to my new full-time job, and Evan had made some reckoning with his parents' silence and found a way to put his feelings about them into words. He told me that he would like to explore joining a support group for the adult children of Holocaust survivors, and I applauded his idea; it sounded like a good next phase in his exploration of both past and future.

As he stood to go, Evan thanked me for our time together, and I told him how much I, too, had valued it. As a parting gift, I suggested he might read something by C. S. Lewis, the author of the Narnia books he had loved as a boy. 'Have you read *A Grief Observed*?' He had not. I can't begin to think of how many people I've recommended this to over the years, and what a valuable resource it has been for me, in work and in life. I assured him there was no wiser or more eloquent voice on the link between loss, fear and memory, and I wished him well.

As he left the session, he fumbled in his jacket pocket and produced a flyer. 'I've joined a choral society here. Over Christmas, we're doing a concert at a church in the City. Would you like to come? I mean, if you have time.' I was glad to hear he was singing again, given our discussion of music and emotion, as well as his parents' great love of music – that was a bond that could transcend death. I said I would try to make it, knowing I probably would not. It isn't that a therapist should never go to an ex-patient's concert, but I thought Evan might just benefit from a good, clean ending; all that needed to be said between us had been said. He had brought his pain and fear to me in our therapeutic space and found a voice to mourn, to protest, and ultimately, to begin to accept his loss and his parents' legacy. It was time for him to transform that into a new way of living, to find his full voice outside my little room, without me. I went to the window and watched him go, stepping briskly, his body braced against the winter cold as he walked into the gathering darkness.

The Children

Imagine if one day you saw a little child – a girl, say, aged five or six – climbing a big oak tree in the family garden. Suddenly, she slips and falls, tumbling to the ground ten feet below. Her leg is badly broken; jagged bone juts through skin as she screams in shock and pain. As you watch, the emergency services arrive, but to your astonishment, the people sent to help do not minister to her leg but scold the child for being stupid, then set about beating her around the open wound.

This gruesome image came to mind as I started reading the bundle of files sent to me by the Irish Redress Board, or IRB, which contained dozens of statements from people seeking compensation for harms done to them in care homes when they were children.[1] A solid thread of abuse and neglect ran through all of them, recurrent stories of children who had already suffered grief and loss and were removed into institutional care, only to experience further, repeated trauma over a number of years.

It was the early 2000s, and I had joined a service that operated as a specialist resource for lawyers seeking experts in different fields, and my profile highlighted my interest in trauma. I turned down a lot of requests for reports on things like work-related accidents, but I was immediately interested by an approach from a firm of London solicitors that was representing people making claims under the IRB scheme. Independent assessments from psychiatrists were needed to assist the Board in awarding damages in each case, much as I'd done years before with the helicopter crash victims. In this high-profile scenario, reports were also being sought in order to deter people from making false claims.

From my work as a forensic psychiatrist and in the family courts, I knew something about how and why children were removed from their parents' care. Generally, it requires a court order after intervention from social services. The state has a duty to protect its youngest citizens if no one else can do so, as in those sad situations where parents may have died, are gravely ill or are incarcerated, and where no other appropriate family member or foster carer is available. In some cases, children are placed in care homes when they are at physical risk, which can happen when their parents or carers actively pose a danger to them or are simply incapable of looking after them and meeting their basic needs.

In Ireland, the large-scale institutionalisation of orphaned, destitute and unwanted children had been a fact of life dating back to the devastation of the nineteenth-century potato famine. Long after other countries, including the UK, had replaced residential homes with foster care in the community, such homes persisted in Ireland because of the close relationship between the country's social policies and the Catholic Church. Nearly all the 'children' I would meet (most of them now in their sixties) had initially been raised in Catholic families; it was, therefore, almost inevitable that they were sent to homes run by different Catholic orders, who also provided schooling. These places were usually staffed by religious professionals (monks, nuns and priests) and lay people associated with a religious order – notably, the Christian Brothers, a charitable group established in 1802 by a philanthropic businessman.[2] The stated mission of this lay order was to provide housing and a Catholic education to disadvantaged minors. However, neither the Brothers nor the clergy who worked in the homes had any formal training in the care, teaching or nurturing of children, especially those who had previously suffered bereavement or violence in the family home.

All this was in my mind as I disembarked from a crowded Tube train in the north London borough fondly known to many Irish immigrants as 'County Kilburn'. I stopped to ask for directions, then made my way down the unfamiliar high street, a busy thoroughfare lined with shopfronts and dotted with restaurants and pubs, searching for the solicitor's office where I was to meet the applicants. As in other personal injury cases, to conduct these interviews I had offered to come to them, because I think it feels easier for people to meet in a space they know already – the layout of the rooms, the exits, the washrooms, some familiar faces.

In preparation for this work for the IRB, I reviewed as many studies as I could about the long-term effects of different kinds of child maltreatment, especially some larger-scale longitudinal studies that followed hundreds of subjects over a long period, comparing their outcomes with those of non-abused children. One study that particularly struck me looked at the response of both abused and non-abused children to their images in a mirror. The abused children reacted with hostility and harsh self-judgement, which did not happen at all with the non-abused ones. A related study showed that abused children found it harder to find the vocabulary to express negative emotions, like anger and distress. These toxic effects on a child's sense of self could limit their capacity to communicate emotionally and create serious difficulties in adolescence and the transition into adulthood.[3]

Importantly, these studies also found that not everyone who experienced maltreatment as a child struggled in adulthood, and that children who were treated well did not necessarily thrive later. The binary, factual legal world abhors such paradoxes. In this case, the IRB's guidance to experts asked for an opinion on precisely how the trauma caused by the care homes resulted in mental health problems later. That kind of approach doesn't allow for the reality that the process of developing a sense of self and other people in childhood is layered and lengthy. But that's what I was there for: to help the Board understand the complexity of trauma and the mind.

A few years after I did this work for the IRB, a landmark study in California on stress reactivity in childhood was published, suggesting that the majority of children seem able to flourish in any environment, whereas about 20 per cent of them need extra attention and nurturance in order to flower. I wish I had been able to use the vivid metaphor the researchers adopted when I made my reports to the IRB; they took the ancient Swedish concept of the *maskrosbarn*, or dandelion child, and the *orkidebarn*, or the orchid child, and used them to frame their hypothesis.[4] Their research showed how children interact organically with their environments, and how different experiences might affect how they express their genes for resilience and vulnerability. Dandelions could thrive in harsh conditions, while orchids required additional help to survive. However, this isn't always true. I knew that going into care and what happens after might act on a child's mind like a hard winter or unforgiving soil – however tough they might be. As an incompetent gardener myself, I was only too aware that any plant will die if you treat it badly enough.

After a few wrong turns, I located the unassuming solicitor's office, where Jacob, the son of the solicitor who had briefed me, ushered me in with a smile. His wife Rebecca offered me tea or coffee. This was a small Jewish family firm, and as I was shown into a meeting room, I found myself wondering if the people I was about to interview had consciously or unconsciously chosen to have legal representation that had no connection to the Catholic

Church. Moments later, I felt embarrassed when Rebecca set a mug of tea before me and commented quietly, 'I noticed you're wearing a cross. That might be upsetting for some?' I quickly tucked my necklace away, thanking her. I also made a mental note that it could be important to talk to people about their current relationship, if any, with faith and the Church.

Bridie was my first interview that day. She was right on time. She hovered in the doorway uncertainly, a slim, neatly attired woman of sixty or so, her short-cropped greying curls framing a squarish face, her eyes intensely blue under thick, straight, black brows. Unsurprisingly, my first impression was of wariness and anxiety, reflected in her gaze and stiff posture. I offered my hand and introduced myself.

'Please. Come in. Have a seat. Thank you for meeting with me, Mrs—'

'Bridie's fine,' she interrupted, with a hint of implied criticism.

'Bridie, then,' I said.

She nodded slightly and sat down, back ramrod straight, an ample black handbag clutched to her torso like a shield.

'Can you tell me how long we'll be, Doctor? It's only I've my daughter waiting, and—' There was something birdlike about her, I thought, a nervy quickness, a tipping of the head to one side when she looked at me, as sparrows do when they alight on my kitchen windowsill. Her voice, with faint remnants of an Irish lilt despite her forty years in London, wasn't exactly chirpy, but it was in a higher register than my own. And much like the sparrows, poised at the edge of her seat she seemed ready to take off at any moment.

I told her we could accommodate her schedule and that I was more than willing to meet a second time, which immediately prompted a brisk, 'No, let's get it over with today. What can I tell you? I mean, it's all in my statement . . . what happened.'

'Yes indeed, Bridie. I've got it here.' I held it up for her to see. 'And I found it very helpful, thank you.' I smiled, hoping to thaw her frost a little. I explained that I'd like to talk a bit about parts of her life that hadn't come up in the statement in order to get a fuller picture.

Her eyes narrowed. 'How do you mean?'

'Could we talk about how things were before the care home? You were' – I consulted the statement – 'aged twelve when you went?'

'Just turned,' she said quickly. 'Our mammy died when I was eleven. She got sick, and it was too much for Da.' The opening lines of her statement, verbatim.

'And before then?'

Bridie frowned. 'She was ill. Had a cough that wouldn't go, for a long time. Years, I think. But she still did everything – looked after us, cooked, cleaned. Her life was' – she searched for the right word – 'hard. But she was always there. And then one day she wasn't.' I waited, saying nothing, sensing she was scrolling back in her mind. After a moment, she went on. 'We didn't even know it had happened. We were at school that day, and then went to my best friend's house next door, like always. Once everyone was at home that night, our da gathered us round and told us.' She stopped, and I wondered if it felt too soon to be talking about something so painful. I picked up on her mention of the best friend, asking her to talk a little more about her.

'There was a gang of us. You know how it is, that age, girls on the school bus together, in class, going around in a pack. Inseparable.' I thought her features softened a little at the memory, and I smiled and nodded, joining her on the common ground of our experience of close, young, female friendships.

'Who else was around when you were growing up?'

She seemed puzzled by the question.

'Tell me about your siblings?'

'Oh. Well, Paddy's the eldest, then Tommy and then Maura. The older boys were twelve and ten when I was born, and Maura was eight. I think . . . I think my mother may have lost a few babies after that. Then I came along, and then my little sister, Kathleen – Katie – and then there was Michael . . .' She ground to a halt again, her face unreadable. Then she added, 'He was the baby. Just five when Mammy died.'

'So, when you were a little girl, who looked after you?' As usual, I was using some of the basic questions from the Adult Attachment Interview, which was so helpful in getting a conversation about childhood memories going. She tilted her head to the side and said she didn't get my meaning.

'Anyone apart from your parents?'

'Um . . . Maura, I guess. Mammy was busy with Michael and looking after the house . . . We were poor, but they cared for us, and we didn't go hungry. And we had parties sometimes, with my father singing – he had such a lovely tenor voice – with everyone drinking and laughing. And once, our Auntie Maeve took us all to the seaside . . .' She stopped. 'Listen to me. I'm getting off the track, aren't I?'

I assured her she wasn't and that anything she could remember was okay to discuss.

'I suppose it was mainly Maura who looked after us little ones. We were so close, Katie and me. Irish twins, you know?'

I hadn't heard that expression before.

'Born that close together. Thirteen months, almost to the day. Anyway, Maura always made sure we got our breakfast, had clean uniforms and were on time for the school bus, until we were old enough to do all that ourselves . . .' She trailed off. 'Is this what you meant?' She wasn't challenging me, just checking, and I nodded, glad she seemed to be actively engaging with the interview, not shying away or blocking my questions.

'Did anything scary happen to you when you were a little girl? Were you ever frightened of your mum – your mammy – or your father?'

Bridie shook her head vigorously. 'Never. Only . . . only when she got sick . . . that was scary. She'd cough and cough all night, and I saw blood on her handkerchief the one time. She tucked it away and said it was nothing. And another time she got very bad, and Father Darragh came, and we all had to be still and quiet in the other room, and then an ambulance took her off to the hospital in the city, but she came back.'

She made eye contact briefly, then looked down at her lap. 'My poor da. He was in bits when she passed. I don't think he ever recovered.' I reflected to her that when children lose one parent by death, they often lose the other to grief – and their elders' distress can be frightening and disorientating.

Bridie agreed. 'It was like . . . Da died, too, in a way. He was a ghost, drifting in and out.' I couldn't help but think of Evan then, and his image of a 'grey ghost' mother, her liveliness desaturated by grief and loss. It sounded like Maura, as the eldest girl, stepped into her mother's role, taking on all the household chores, even though she could only have been in her teens. The father and older brothers were out all the time, working on their smallholding. I knew that farming was a hard life at the best of times. 'It was a different house,' Bridie told me. 'Katie and I would look after little Michael, but he'd be crying for Mammy, and that made us cry, too, and . . . everything seemed so . . . wrong, you know?' She sniffed and dropped her head, but I saw her eyes were full.

'And then what happened?' I said, after a pause.

She didn't look up. 'Well, Father Darragh kept dropping by in the evenings to talk to Da in the sitting room with the door shut, and then a woman came along from I don't know where and told us three younger ones we were going to have to go live someplace else because our daddy couldn't afford to look after us anymore.'

Sorrow cracked her voice, and she busied herself with opening her handbag to rummage around for a handkerchief while she tried to regain her composure.

I needed to help her contain her distress so she could continue and get through the rest of the interview. Quietly, leaning slightly towards her, I said, 'I can see how close to the surface this memory is. Time doesn't change it, does it?'

She shook her head, wiping her eyes.

'Bridie,' I told her, 'you've really helped me understand what life was like before you went to St Columba's. Thank you.'

I picked up her statement, turning a few pages. 'And you've gone into a lot of detail in your statement, so we don't have to go over that again . . . but I would like to ask something. How do you think you survived all that time – how many years was it?' I knew the answer; I had the dates in front of me, but getting back to facts and away from emotions would help her settle.

She answered without hesitation. 'Five years and three months, all in. Felt ten times longer. After Katie turned sixteen, they moved us to a house for older girls in Dublin. We'd do cleaning jobs, work in pubs to earn a little money. And as soon as we could save up, we came away to London.'

I was thinking about the people I had met who had fled their homelands due to crises like famine and war. For them, going home would never be an option, but many spent a lifetime mourning that loss. I made a note to come back to her feelings about leaving Ireland, but for now, I needed to stay with her statement.

'You say here that when you and Katie went to St Columba's, your younger brother was placed in a different care home?'

She nodded. 'For little ones, yes. I think it was run by the Sisters of Mary, too. It wasn't near us; further south, down Waterford way.'

'Can you just talk a bit about how you and Katie coped in the home?'

That was easy for her to answer. 'We had each other.' She didn't elaborate, and I nodded, making a note. In other cases, I'd seen how people who had siblings in care with them appeared to have some psychological advantage, a supportive grounding that children who were alone did not. Bridie added that Maura would visit whenever possible, saying that had helped too.

'But the nuns discouraged visitors, really. And when she did come, it took three buses, and she had to work all hours, so . . .'

'Did you tell Maura what it was like there?'

'Sure we did, and she was shocked and said she'd speak to Father D. straight away. But nothing changed. And then Maura met Kevin, and they got married, and then she had a baby of her own, so she couldn't come any more.' I could hear that adult Bridie understood why, but I imagined how sad and abandoned the young Bridie might have felt. She and Katie effectively lost touch with all their family in the ensuing years; her statement explained, again from an adult's perspective, that the farm was failing, along with her father's health, and there were no resources to make travel possible.

'Are you in contact with Maura today? Do you ever see her?'

Bridie shook her head. 'She and Kevin moved to Donegal after they had a family, I think. I tried to find out where, but . . . we lost touch.' Her tone seemed to close down any further questions. Today, probably nobody falls out of touch unless they want to, but back then, it was more common – particularly in hard-working, poorer families who did not have the time (and in some cases, the literacy) to exchange letters. I felt curious as to why Bridie had not got someone to help trace her sister, especially when technology had made it easier, but maybe that felt impossible, a bridge too far. I also considered whether she felt Maura had forgotten her, and how much that might have hurt, but I decided not to pursue this.

Had Bridie ever returned to Ireland? A quick shake of the head and a curt 'No' again seemed to close off my line of enquiry and suggested she might have felt abandoned by her country as much as her family, even if she'd been the one to leave. When you've gone through something so dreadful, and you're able to get away from the place where it happened, why would you return? Avoidance of reminders of the past is a common way to cope with trauma; forgetting is about not looking back.

We returned to our conversation about life in the home. 'Katie and I made friends with a few other girls and made our own gang,' Bridie continued. 'That helped. We'd protect each other from the bigger girls. There were some awful bullies there, and the Sisters seemed to encourage them. And they would hit us, too, for the least little thing. They boxed our ears, slapped us – I said all this in there.' She gestured towards the statement in my hands. 'I knew it was wrong. Our mammy and daddy never hit any of us. So me and Katie, I guess we held on by dreaming of how things would be someday when we got out. How we'd go far away, and life would be good.'

She reminded me so much of some of the incarcerated people I'd worked with, who described similar coping mechanisms, like finding a protective friendship group, if they could, and building a future life in their mind's eye. Although she and her sister were not in the care home because of any fault or crime of theirs, it was a version of prison that Bridie was describing.

There was much more in her statement, but I did not ask her to rehearse it further. It made for grim reading. Adolescence is a turbulent time for most of us, but for Bridie and her peers, it seemed like an endurance test conducted in an environment of neglectful sadism. Aside from the slaps and punches, she had testified to constant verbal cruelty, including being told daily that she was stupid, ugly, sinful and lazy, which was just as wounding. In those teen years, when her young body was changing and developing, she was also coping with undernourishment that bordered on starvation rations, by the sound of it.

Who were her tormentors, and what had happened to them in their childhoods – or later – to make them so full of anger and shame that they had to punish their vulnerable charges in this way? Yeats's early poem about fairies taking children from their parents ('Come away, O human child!') and promising to protect them from the troubles of the adult world – a kind of fantastical inversion of Bridie's harrowing experience – leaped to my mind, with its eerie, rhythmic refrain, '. . . the world's more full of weeping than you can understand'.[5] I sat there, gazing down at the litany of hurt on the pages before me, until finally I could muster some words, however inadequate, to reflect my response.

'It all sounds very hard, Bridie. So . . . unfair.'

'Ah well,' Bridie shot back, 'life's not fair, is it now?'

'What was the worst thing, looking back on that time, would you say?'

She was not stumped by the question, as some people are. 'The meanness of their words. Sisters of Mary – you're joking. Those women could be so cruel, on purpose. One morning, very early, one of them came to our room and shook my friend in the next bed awake, only to tell her that her father had passed. I'll never forget it. Mary Louise was in shock, and the nun leaned down close and hissed at her, "No one wants you now."'

Injury upon injury again. No chance for the bone to set.

'Is that enough?' Bridie asked, with a glance at the clock on the wall.

'You tell me. Do you want to add anything?'

'I guess . . . I couldn't always protect Katie. That was bad.'

She sighed then, her breath ragged. Clearly, this brought up an extra-painful memory that needed extra effort to share with me.

'Go on.' I sat very still, wanting her to know I was ready to hear anything she had to tell me.

'So, they made us line up when the priest came to give us a blessing, and we'd all have to sit on his knee. I wasn't stupid. I knew what was what. We heard stories from some of the older girls. When it came to my turn, after he made the sign of the cross on my forehead, I tried to get up straight away, but he tightened his grip and his hand moved down . . . to my skirt, you know?'

I nodded, indicating she didn't have to be more specific. As she spoke, I noticed her hands had balled up into fists by her side, and her accent, which was quite soft, tempered by decades away from Ireland, began to thicken as she recalled what came next.

'And then?'

'I jumped up and gave him a swift kick, so I did. Right there, know what I mean?' She brought her hands together, and her right fist punched hard into her left palm. Thwack. I kept my face neutral, though there was something shocking about this sudden, vivid glimpse of young Bridie. As well as the accent, the quality of her language shifted, becoming briefly rougher and less formal.

'They gave me an awful hiding for that one, don't you know, but I was used to it. I was fifteen, and I wasn't going to let some dirty bugger touch me up, even if he was a priest. Not me.'

I was intrigued that she hadn't mentioned this man in her statement, unless I'd missed it somehow, which was unlikely. 'And Katie?' I asked, thinking of how she'd begun this sorry story.

'I told her to do the same as me – fight back. She did. Then they beat her black and blue.' Bridie's face clouded. 'I felt terrible. She was a little slip of a thing, not strong, and she was in bits after that, crying and crying for our mammy all night. That was my doing. My fault.'

She closed her eyes, and I held myself still and silent until she composed herself and reached into her bag again for her handkerchief. As she did so, the movement caused a chain round her neck to move slightly, and I realised she was wearing a small gold cross, almost identical to mine. Given what she'd just told me, I felt anxious asking about it, but I went ahead. 'Bridie, do you have any religious identity today? Do you have a connection with the Catholic Church?'

She touched her necklace briefly, perhaps realising I'd seen it. 'I didn't for ages,' she said. 'But once we got settled here, there was a church down the road, Immaculate Heart. I could hear the bells for Mass on Sunday, and—' She shrugged, a little awkward. 'One Sunday, I went. On my own. And it was all right. Reminded me of when I was small, of my mammy, and . . . I liked it. The familiar words. I met some kind people – including Joe, my husband. He's not Irish, though. He's Polish.' Her face relaxed. 'He's a good man.'

I was not surprised that she had still been able to see a church as a place of safety and connection, even after all she'd been through in a Catholic care home. My experience as a trauma therapist has taught me not to make assumptions about what stays with people. I had learned a lot about that state of mind that Keats called our 'negative capability',[6] a beneficial embrace of uncertainty outside of facts and logic – which for me implies that both bad and good things can co-exist. Bridie's experience exemplified this and made me think she was a dandelion child whose positive memories of early childhood helped her distinguish people you could trust from those you couldn't. Making new secure attachments may have helped her let go of the attachment to Ireland that was now so entwined with psychological pain. So going to church and finding new friends there as an adult activated a positive memory of feeling safe and connected as a child. It gave her somewhere she could belong to that was familiar, a place of kindred souls.

'Did you and Joe marry there?'

She nodded, and her face softened. 'And then we had our three girls. I made another little gang of girls again.' Her smile reached her eyes. 'I'm only here with you because of them, you know. And Katie.'

'How so?'

'My girls know a bit about my past. I told them some things once they were old enough. Only some things, mind you, but enough. Our eldest, Colleen, is a paralegal now, and when she heard about the Board, she said I ought to go along and make a claim. I wasn't sure, but then I thought about Katie. I would do it for her.'

'For her?' I echoed, encouraging her to say more. None of this was in her statement either.

After they came to live in England, Katie struggled with mood swings and anxiety, and sometimes she 'took too much drink', as Bridie put it. About ten years after they settled in London, and Bridie and Joe had married, Katie met a nice Irish lad. Barry's family had recently emigrated to Australia, and soon he proposed to Katie, asking her to go with him so they could live with his people.

'They'd start a new life. Maybe get a farm. That was the idea.'
Bridie swallowed hard. 'Katie seemed excited, a little lighter. She
wanted that new life so much, so I wanted it for her, too, I did. But
I missed her dreadfully when she was gone. Felt like I'd lost an
arm or a leg, honestly. We'd always been together, you see. But I
encouraged her – "Go on, do it."'

The sisters wrote letters and did their best to stay connected
as they settled into married life and began having children. About
six years after the move to Australia, Katie's letters stopped. After
three months of silence, Barry wrote to tell Bridie that her sister
had died in a car accident. He was left with two young ones to
raise – 'a niece and nephew I've never met'.

More motherless children. I could see what anguish this caused
her, too. Her brow furrowed, and her voice was quiet, almost as if
she were talking to herself. 'Was it an accident? Was she drunk?
Or maybe she was so down . . . Did she drive off the road one night
on purpose? I'll never know. We lost touch with Barry and the
kids after that. And that made my mind up about this application.'

'Because?'

'I told my girls, "If I get any money, we'll use it to go down
there and find Katie's grave." Colleen said she'd come with, if we
can manage it. We want to meet her children, too, if we can.'

I had half a mind to write a cheque myself to this woman who
had experienced so much loss yet been so heartful and fierce in
her support and love of her sister. Bridie had somehow stayed
connected to hope in the face of much grief and fear, and after all
she'd been through, this brave soul was prepared to sit and talk to
strangers like me about some genuinely awful memories. I felt in
awe of her fortitude.

I had one final question. 'Bridie, have you ever talked with
someone like me before? A psychiatrist or a therapist, maybe? You

mentioned Katie struggled with anxiety and depression, but what about you?'

Bridie made a gesture with her hand, swatting the question away. 'Oh no. No need. I mean, I have people around me who love me – Joe and the girls – and I know my parents always loved me . . . but then Katie had that, too. I must be made different from her. It's like she wilted, and I grew stronger.' The dandelion and the orchid. Even a lovely husband and a new life had not been enough to sustain poor Katie after all she had endured in child-hood. It did occur to me that Bridie could have been deflecting from her own state of mind, but I decided to take her at face value.

I had everything I needed. All that was left was to ask, as I always do, if there was anything she wanted to ask *me* before we parted. She hesitated. 'Um . . . yes. About the priest that time – really now, there was nothing to it, Doctor. Nothing bad . . . that is, nothing *sexual* happened. Could you please not put that in your report? They might think it was my fault. Don't put that in.' I could almost see the young Bridie before me, a fierce and deter-mined girl facing down strange adults who controlled her fate. I heard the shame in her words and imagined the nuns may have told their charges that any overtures from the priest were their fault – 'You asked for it.' I reassured Bridie I would say nothing of the incident, and she thanked me.

She could not know, as I did, that there was some guidance from the IRB about claims of sexual abuse, which meant extra compensation if people reported it. I found this approach baffling since it diminished the impact of other kinds of abuse, like neglect and physical violence, which seemed to have happened to these children almost daily. The idea that you might get a few thousand extra euros if you had been abused sexually gave that type of harm a strange status, as if there were a competitive hierarchy of cruelty to children. In my reports to the IRB, I made a point of emphasising that all kinds of abuse and neglect need to be counted equally seriously, not least because they can often happen in tandem and exacerbate each other.

'Oh – one more thing,' said Bridie. 'I think my brother, Michael, is part of all this, too . . .' She made a gesture encompassing the meeting room, the door and the solicitor outside. 'Colleen only just told me. We don't talk, Michael and I – not any more. He . . . well, some bad things happened back in the day, when he first came over. I tried. My husband tried, even gave him a bit of work. Then he stole some money from us, for drugs, I think . . .' She looked embarrassed. 'But he always had a soft spot for our Colleen, and she does her best to keep in contact. She told him about the application. Encouraged him to do it as well.'

I thought this daughter sounded as impressive and committed to the family as her mother – and now I was curious about the relationship between Bridie and her youngest brother. But that was beyond my scope. We were out of time.

At the door, she turned back. 'Doctor? If you see Michael, maybe you could send him my love?'

'I'm sorry, Bridie, I'm not sure if he's one of the people I'm due to meet.' Other experts besides me were doing reports for this law firm. However, I was aware that some claimants were related to each other, and afterwards, I learned that one legacy of the IRB was that it sometimes promoted reunions in families, if they were estranged. I told Bridie I'd do what I could.

She briefly put her hand to her face, as if to mask some emotion. 'Just give him my love – if you see him.'

'Of course I will.'

———

As part of the preparation for writing my reports, I reviewed the data on the long-term effects of adverse childhood experiences (ACEs), derived from a large-scale study that American researchers had conducted in the late 1990s.[7] The list of ACEs includes all the grimly familiar types of childhood maltreatment, categorised as abuse, neglect and 'household challenges', and encompassing physical, emotional and sexual abuse, exposure to domestic violence and parental addiction. ACEs include experiences that don't involve physical touching of any kind – such as verbal and emotional cruelty, early bereavement, extremes of poverty, hostility in the home and having a parent in prison. Subjects were asked about their early lives to calculate an ACE 'score', which became a method to help insurance companies and public-health experts better understand who was most likely to use health-care services. Their work involved 17,000 people, and among their findings was the stark fact that around 10 per cent of the general population in the US have experienced four or more ACEs, and they are most at risk of poor physical and mental health in adulthood. These results have been replicated again and again in different countries since then: more recently, a 2023 Canadian meta-analysis of children from eighteen countries showed that nearly 15 per cent had experienced four or more ACEs, skewing higher for children in residential care.[8]

I was anticipating that the IRB applicants would have lots of ACEs, and sure enough, aged only twelve, Bridie had experienced poverty, the early death of her mother and the associated family breakdown. Sadly, like other young people in care, her ACE score rose, as the staff at St Columba's added at least four additional ACEs to her tally. I would take care to distinguish the impact of the multiple abuses in the care home from the emotional impact of the loss of her sister in adulthood. Without mentioning the abusive priest, I would note how Bridie had also been affected by her sister's experience of being beaten. Although I couldn't make a formal diagnosis of any medical condition, I reckoned this summary of her suffering would legitimately place Bridie in a category of trauma that would merit reasonable compensation, and I hoped it would be enough for her and her daughter to go to Australia.

I had heard what Bridie had said about her family support system and her faith. She embodied grace under pressure and seemed quite well in herself. Still, I decided to put a recommendation in my report that she might need some therapeutic support in the future, especially if she was going to make that long journey to say goodbye to her sister. By now, I knew enough about trauma's self-determining and often circuitous timeline to conceive that new stressors may open old wounds, even for people like Bridie, who appeared so resilient and strong.

I checked with Jacob and learned that her brother Michael was, in fact, scheduled to come in to see me about two weeks later. But it would be over a month until we met. He failed to attend two appointments, cancelling at the last minute. This pattern was so familiar to me from my days in the trauma clinics.

Unlike his sister's statement, Michael's was sparse. It ran to only two pages, and the language was bare, matter-of-fact. He confirmed the time he'd spent in care – nearly eleven years – and that there had been 'physical and verbal abuse', but offered little detail. There was a brief mention of him leaving Ireland as soon as he turned sixteen, and the line 'Found out my dad was dead and two of my sisters were in London, so I went there', but he had not given their names or said anything more about them. He mentioned his marriage in 1980, two children and a divorce by 1990, all couched in the linguistic equivalent of 'Move on, there's nothing to see here.'

I'd seen plenty of data about the importance of timing when we think about childhood trauma and its impacts.[9] Michael's experiences looked superficially like his sisters': taken into care after the death of his mother, placement in Catholic care homes, and ultimately leaving Ireland to come to the UK. But when the initial, devastating break occurred, Michael, 'the baby of the family', had only just turned five – so much younger than his sisters Bridie and Katie, then aged twelve and eleven.

I was prepared for the possibility that his attachment system might be more disrupted and disorganised than theirs. At five, he did not have the brain capacity to formulate happy memories of going to church with his mother or to enjoy and encode the echo of Da's fine Irish tenor voice around the house. At that age, children also have a less developed ability to make sense of distressing events and emotions, having not reached the stage where they can fully 'mentalise', which is the imaginative capacity to think about our minds and those of others. If we have a secure attachment, we develop good mentalising skills and gain what British psychoanalyst Professor Peter Fonagy has called 'epistemic trust'.[10] This is not just ordinary reliance; it is trust that the person you are relating to sees you and knows you as an individual. If you tell them how you are feeling, they say, 'I know what you mean.' And if they tell you a place is safe, you believe them. Only with mentalising and epistemic trust is it possible to make sense of the world, your place in it and how to relate to other people using the language of emotions.

Exposure to trauma in the first five years of life can lead to a lack of a secure sense of self and a tendency to dissociate from feelings of any kind of pain. Small children naturally dissociate because they lack the brain connections to process distress; those apparently happy children photographed playing in the rubble of a bombed-out city are almost certainly in a dissociated state because there is no one around to help them feel safe. However, chronic dissociation for weeks and months is not good for a child, especially for their right brain, which they need in order to grow a social mind.[11] As the traumatised child develops and gains agency in adolescence, if they continue to use dissociation as a means of managing distress, they will struggle to make the peer relationships that are so important to adult security. They may also have difficulty in relating to their bodies and experience suicidal ideation and the compulsion to hurt themselves.

Michael's story was typical of how so many young, traumatised people escape from pain, seeking rescue in the grape and the poppy. Addiction becomes the solution. In his statement, Michael spoke of having been 'in the wars with drugs and booze' since his late teens, including a heroin habit, which he stated was no longer active. Based on the patchy work history in his file, I guessed alcohol might still be a live issue, possibly affecting his capacity to stay in consistent employment.

He was nearly half an hour late for our appointment. When he did arrive, my first thought was how little he resembled his sister. If she was a sparrow, he was bearlike, lumbering and clumsy, his long, grey-black hair matted and unkempt. His voice was low and rough in quality, and he almost knocked over the chair I'd set near the door for him. With his history, it was possible he was drunk. But as we began to talk, his words seemed clear enough, and I guessed he was just anxious in a different way to his sister. His language was respectful, like hers, but he addressed me first as 'Madam', then 'Doctor', then 'Mrs', as if he were casting about in order to work out which was correct. I let him do whatever he liked. I would need to be warm, careful, gentle and slow so as to make the room and my intentions seem safe.

I asked if he would like water or tea and enquired how far he'd come, adding that I was glad he could make it and could imagine how hard this must be for him. I repeated what I had told Bridie, giving my usual disclaimer, which is intended to help people unclench a little at the start of an interview like this: 'I'm not here to make you go over everything you described in your statement.' He gazed at me, uncertain what to make of that. 'I thought they said . . . All right, whatever you say, Mrs.' I explained that I'd like to know a bit about his life before the home and something about its impact on him as he saw it. He frowned but did not comment. I also reiterated that he was free to step out if he needed a break, and we did not have to get through everything today.

'I'd rather get it over with,' he said, echoing his sister's words, although his Irish accent was far more pronounced than hers. Unzipping his worn, leather flyer-style jacket in one irritable movement, he threw it over the back of the chair, then settled low in the seat, his shoulders slightly hunched. I gazed at him for a moment, at his widow's-peak hairline and heavy, dark brows. He must have been quite handsome once. Although I knew he was several years her junior, he seemed older than his sister, with all the tell-tale signs of a habitual drinker, including a reddened face, with a fretwork of broken veins spreading across his nose and cheeks.

He jerked around in the chair, fishing in a jacket pocket for something, then pulled out a crumpled packet of cigarettes and a lighter. 'Smoke in here?' I was tempted to let him because it might put him at ease, but there was a prominent 'No Smoking' sign in reception, so I had to apologise; it was not allowed. But he could have cigarette breaks whenever he needed them.

He shrugged, as if this wasn't a big deal, then held on to the packet anyway – something to fidget with as we talked. 'I'm off to a smoking meeting soon anyway,' he said.

'A smoking meeting?' I was confused.

'AA. Al-co-holics An-ony-mous,' he articulated. 'They let you smoke in meetings at the community centre round the corner, long as you clean the ashtrays at the end. Free cuppa and decent biscuits, too. I'm off there before I start my shift tonight.'

'AA,' I repeated. 'That sounds important. How long have you been sober?'

He briefly raised his eyes to the ceiling, lips moving as he calculated, then told me it was forty-eight days. No, forty-seven. 'My name's Mick, and I'm an alkie all right.' He smiled then, a lopsided, mirthless grimace that briefly revealed a set of terrible teeth, chipped at the front and yellowed by nicotine.

'Would you prefer I call you Mick rather than Michael?' I was thinking of Bridie using his given name but knew it was too early to mention her. This session was about him.

He nodded. 'I've always been Mick, ever since I was sent to that fucking place.' I was immediately interested in the fact that he had introduced the subject of the care home so early in the interview. The obscenity was casual, almost unconscious, and I noted that he did not check out my response, as if he was barely aware of me and perhaps didn't care how it might land. People use swear words in so many ways – sometimes for emphasis, sometimes to intimidate – but in this case I thought it signalled how Mick was already dissociating from this interview because it was so painful to talk about what had happened to him in that 'fucking place'. I decide not to comment and opted instead to stay with Alcoholics Anonymous, asking if he could tell me a little more about it. He said this was his third time in the programme, his 'th' sounding like a 't'. He flashed the briefest of smiles. 'Third time lucky, eh?' He seemed to want some confirmation of that from me, but I simply commented that I'd heard good things about AA and knew that many people found it helpful.

'There's plenty that don't make it,' he said. 'Go along only for the tea and biccies . . . Get there late, leave early. That's not the way to do it. You have to want it. That's what Carmen says.' Then he added, 'That's my sponsor this time, Carmen. An old-timer. Sober fifty years or something. She's good . . . She's helping me. Trying, anyway.' That was promising, not least because he had forged a relationship with someone he valued. I had heard the saying, 'The opposite of addiction isn't sobriety, it's connection.' Addiction is lonely, and anyone who reaches out to help a person with a substance use disorder punctures their isolation. The fact that his sponsor was an older female might also give Mick a sense of being held in mind like a child by a mother. He had lived much of his life unmothered, even 'unmoored', a little boy carried away from a loving family into cold and hostile waters, with no one to turn to for aid. Desperately sad.

I moved on with the interview, asking him what he did for work; his file listed a few things, like gardener, builder, bartender and other odd jobs – lots of chopping and changing. 'Minicabbing at the minute,' he said, 'but the problem is . . .' Throughout the interview, that seemed to be a favourite refrain, creating the impression of someone without agency to whom things just happened. He then embarked on an involved story about some 'fucking morons' he'd picked up in the City last night who'd vomited on the back seat and then jumped out without paying. He'd tried to run them down in the street, but they got away. All this came across as a rant, and I was aware of feeling a twitch of annoyance with him, which I certainly hadn't had with his sister. I considered how angry this man might be and whether he was already finding the interview too stressful.

He was on to another complaint about how 'bloody hard it was to get a decent job with a record'. That was a problem I knew about, and I told him so, glad to get a word in. I was familiar with the frustration that came with the seemingly indelible identity of 'offender'. He went on to tell me about his winding path in and out of prison over the last several years – short stints, mainly for assaults on other drunk men, a bit of criminal damage 'and some other stuff'. When he paused there, looking uncomfortable, I sensed there was something he didn't want to tell me. The thought came that there might have been some drinking and fighting at home, possibly related to the break-up of his marriage, but I would not ask point blank if he had a history of intimate partner violence. I was treading with care.

As if our minds were in synch, he segued into talking about his wife Caroline, their two children (now in their teens) and the divorce. 'How did that come about?' I asked, as gently as I could. Mick hung his head, fiddling with his cigarette packet, flipping the lid up and down. 'Ah, it was the drink. Both of us, really, but me mostly . . . and there'd be rows about money, too. Then the fights got kinda out of control, you could say. The kids were only little, and . . . they saw. And I . . . I . . .' He stopped. 'Can I have a quick fag now?'

'Of course,' I said. I told him to take all the time he needed. When he returned, he had a glass of water in his hand, and he took a long drink, draining it before he spoke. He told me quickly, without making eye contact, that 'the problem was' that he'd hit Caroline 'the one time', so she'd taken the kids and left, went to her mum's and never returned. He'd tried to apologise and begged to see the children at least, but she got a restraining order against him and filed for divorce with full custody. The complaining tone was gone now, replaced by something much softer and sadder, as he talked about how he hadn't seen his children again for nearly two years. When he did, it was only for a supervised visit, after he'd proven that he had some months of sobriety under his belt, and it went badly. 'Almost like they didn't know me,' he said. He looked down at his lap so I couldn't see his eyes.

'This is upsetting for you to talk about,' I offered, not knowing what else to say.

'I don't want to be that kind of man, don't you see?' His voice was suddenly loud, almost a shout, making me start a little, but it was so clearly pain, not anger, that he was expressing. 'Jesus. After what they did to me, how could I be so crap, turn into someone who hits women, frightens little children? That fucking place, those fucking people, I blame them, you know? But then . . . I start to worry if I'm just like them . . . Do you think I am, Doctor?'

This was not a therapy session; it was a formal interview for legal purposes. Yet here was someone in distress who needed a therapeutic approach. So I told Mick that I had met many men with similar experiences to his, and I knew that managing cruel and angry feelings is one of the most challenging things for trauma survivors to do. 'Sometimes it's easier to avoid bad feelings and say, "Fuck it,"' I said, deliberately mirroring his language. 'I wonder if you become hyperalert when faced with any sort of conflict or threat, and maybe you use hostility to protect yourself? I'm just guessing here.'

He stared at me, nodding. 'They said something like that to me in prison. People talk about that in AA, too. Old behaviour isn't old behaviour if you're still doing it, Carmen says.'

I steered him back to the job at hand. 'I know you were in care for a long time, first in a home for smaller boys and then in another one when you got older. How did they compare?'

He sighed heavily. 'They were both shite in different ways.' He talked a little about how nuns ran the first home, and how some of them were okay, but he struggled with reading and writing, and the teachers kept telling him off for being stupid and lazy. He figured they were right; it was only much later that he learned he had dyslexia. How often I'd heard that – and what a tragedy it is when adults repeatedly characterise children with learning difficulties in negative terms, which becomes their identity and can be hard to shake off.

Mick had missed his family desperately and didn't understand why he was not with them. Nobody ever explained it. His older sister and father came to visit him only once or twice, but then (as for Bridie) the visits had petered out. After a while, Mick said he began to believe what the nuns often told him: his family had sent him away because he was a bad boy and no one wanted him.

He said that he did not recall much physical abuse of the younger boys, but he remembered plenty of verbal abuse and punishment. The Sisters regularly sent them to bed without supper or handed out extra chores if they were 'naughty', including some that were demanding for small children. He was constantly getting in trouble; when I asked for an example, he recalled one time when he didn't make his bed, and one of the Sisters struck him around the head so hard his ears were ringing for days. 'Who woulda thought a tiny old nun had so much power in her fists?' He shook his head. Who indeed? I thought, but did not say. I noted, too, how he seemed not to notice that this memory of being physically assaulted contradicted what he had said earlier about 'not much physical abuse'.

At the age of eleven, he was moved on to a care home for older boys. I noticed his language became harsher as he described the transition to secondary school, a shift which many children struggle with even in ordinary circumstances, as it involves becoming a small young fish in a big unknown pond. This can be offset by coming from a loving home and having supportive teachers, but

Mick had neither. And as a new arrival, he was vulnerable prey for the older boys.

'There was this gang of big lads who went around together, picking on anyone that looked at them sideways, giving people nicknames,' said Mick. 'I was a little fellow, a late starter, and they called me Runt. From day one, they teased me about how I looked, how I walked and what I said. If I tried to protest, they would chant "Runt, Runt, Runt" and laugh at me like fucking hyenas.' He recalled that the taunts were so constant and belittling that, combined with the various physical torments the bullies devised, a couple of Mick's classmates – he did not call them friends – ran away. He never tried that, though, he added quickly, because 'I didn't know where home was'. There was no self-pity in those stark words, just a deadened quality, a flatness.

I knew worse lay in store, because it was in the statement. The Christian Brothers ran the new home and secondary school. Mick had testified that the Brothers were as verbally abusive as the nuns had been, telling him he was wicked and was going to hell, but now there was an element of random and quite brutal violence against him and the other boys, which he had not known before, and which he naturally found terrifying. His statement included a brief mention of his hatred for a Brother Matthew. When the solicitor had asked what he did, Mick had stated, without embellishment, 'He battered our faces in.' The constant abuse, the lack of food and friendship, the bitter cold in unheated dormitories and the continued difficulty he had with schoolwork made life increasingly wretched.

Then, when he was thirteen, a new teacher, Brother Nicholas, came to the school. He oversaw Mick's dormitory. At first, he seemed a good sort, talking to the boys about the rugby and so on. But soon enough, he began to play them off against each other, encouraging the bullies and even joining them in persecuting the younger ones. He made positive relationships with the smaller, weaker boys, as if he were going to protect them, then used this to coerce them into sexual acts with him. Or he would just 'fly into a rage' and assault the boys with his hands and with his penis. Mick's statement said that he remembered hearing crying and screaming in the night. When he got up the courage to tell a senior teacher about it, the man accused him of lying, sent him to his room and excluded him from meals for two days.

I asked Mick if he knew what happened to this Brother Nicholas. 'He went to prison in the end, I think. Saw something in the papers once. Years later.' He shrugged, his tone indifferent. I had to ask the next question: 'Did he assault you, too, Mick? You don't mention it in your statement.'

He turned away, and in profile, I glimpsed something of his sister, a certain set to the jaw. His dismissive answer recalled hers, too: 'Nothing like that. Nothing dirty.' His voice shook a little. 'He did hurt me, though. Cut my heart out, I tell ya. And the thing is, Doctor, there was no need! No need. I never said anything about it in there,' he added, with a wave at his legal statement. 'Only because . . . nobody would believe it, nobody.' I felt clueless as to what he meant. Bitterness and rage seemed to flood the space between us, and I made no reply, instead sitting as still as a statue and waiting. It felt so important not to push him to talk about whatever it was, if he could not or did not want to.

It seemed as if several minutes passed. Mick took out a cigarette, then put it away again. He avoided my eyes. Eventually, I said quietly, 'You can say as much or as little as you like, Mick. Take your time.'

Scraping his chair back, he stood up and walked towards the window, where he stood for a minute, hands in pockets, looking out at the unremarkable view of bustling shoppers, schoolchildren clustered at a bus stop with their backpacks, and black cabs, red buses and cars shuffling along the congested road, their beeps and growls punctuated by the odd siren on the banal soundtrack of inner-city life. I saw his shoulders move and thought he was crying, then realised he was taking a few slow, deep breaths. I had the impression he was counting to ten. I thought, Oh, someone has shown him how to self-soothe. That was a good thing. 'All right, then,' he said, after a minute or two, turning to face me.

I made a gesture, inviting him to sit opposite me again, and he complied. 'Ever seen a fox cub?' he began. I almost laughed; it was such an unexpected conversational turn. He didn't wait for my answer. 'Funny little things, pointy ears too big for their faces. But sweet.' He held his hands a foot or so apart. 'Mine was about so big, maybe the size of a tomcat or something.' He said he must have been around twelve or thirteen when he found her. Her paw was injured – perhaps she'd been left behind by her mother and the other cubs – and she was cowering in a shallow indentation behind a wooden outbuilding. A female. He leaned towards me then, animated for the first time. 'Look here, here she is.' He tugged up the sleeve of his worn jumper and showed me a tattoo on his left forearm: a stylised image of a fox, fluffy tail raised like a pennant, beautifully rendered.

He painted a picture with his words, becoming markedly more eloquent and less obscene as he told me his story. One autumn morning, while doing some garden chores, there was a rustling noise, and he found the animal hidden among some dead leaves. She froze, 'scared out of her wits', but he held out his hand and spoke softly, telling her not to worry, he would never harm her. Then, digging in a pocket, he found a little breadcrumb or something and offered it, staying still, 'not even breathing', until she darted forward and took it, then ran off into the trees.

The next day, he went out again and was delighted when she came snuffling towards him across the garden, wary but not so fearful. This time, she let him pet her between the ears with the tip of his index finger, and then she ran off again. After a week or two of this, she let him pick her up and hold her awhile, warm at his chest. 'We were friends, you see,' Mick said, with a brief smile, and I felt his guard was finally down. He was letting me in.

He named the little creature Ava, 'like the film star'. In class, he began sketching drawings of her in the back of his textbooks or the margin of a copybook, shading and refining until he got her exact likeness. I was rapt, caught up in the story, curious if the tattoo on his arm was also his artistic creation. I might ask him later. I knew this story was important and such a good sign of resilience. Even in oppressive conditions, after years of institutional life, young Mick had retained the capacity to love and nurture, enough to make this touching connection. Ava must have been a great comfort to him, and I said so.

I was not ready for the twist. Brother Nicholas found some of his drawings one day, and 'the old bastard put two and two together', Mick told me flatly. One day, he followed young Mick down to the bottom of the garden, where he was holding Ava, coaxing her to try some cheese. Brother Nicholas shouted something like, 'Filthy! Give me that filthy thing!' and he ripped the fox cub out of Mick's hands, flinging her to the ground. She lay there, stunned, twitching. Grabbing a stone, the man brought it down with all his might on Ava's head. 'Smashed her brains in,' Mick said matter-of-factly. 'Brains and fur everywhere.' Then Brother Nicholas took him by the collar and hauled him indoors for a beating.

I felt stunned, rendered speechless by this cruel demonstration of useless spite; the world truly is more full of weeping than we can understand. What could I say? Mick was also silent, his head bent, his face drained of colour and his arms folded protectively across his chest. Eventually, I made eye contact with him, which helped me communicate my feelings.

'Mick, I don't know what to say about this man's cruelty. I believe your story, and I can also see why you left it out of your statement, although it's like all the other violence, isn't it? Cruel and pointless and so sad. Can I ask how you are feeling after saying it out loud to me?'

Mick lifted his shoulders. 'I don't know. Empty.' He sighed. 'For years, I didn't want to think about it. When I was out there, using and boozing, you know? Maybe I didn't even remember Ava for a long time. Blocked her out, is that what you say? I only did this', he said, gesturing to the tattoo on his arm, 'the last time I got out of prison. Everyone inside has tatts with their girl's name or their gang, maybe their footie team. I had her. At least I did, once.'

I felt like the door between us was open, and I decided to step through. 'Mick, I want to ask what you think has helped you make it this far. It hasn't been easy. Is there anything that gave you hope, that you had faith in?'

'Not the Church, for Christ's sake.' The line was bleakly funny, but he was deadly serious. 'I've been thinking about this with Carmen, because AA is meant to be . . . not religious exactly, but spiritual, you know?' I nodded. 'Normally, that kinda talk makes me want to run a mile. "Faith means nothing," I told her. "Take it from me." But she's got me thinking about this in another way, like you can have faith in your kids, or people in the meetings or family, like Bridie and Joe, my sister and her man. They've been good to me in the past. But they didn't want to know me after the divorce and all.'

'Are you sure of that?' I ventured.

He looked uncomfortable. 'Maybe. I mean, my niece Colleen is in touch. That's their eldest. I see her sometimes. She's why I'm here, actually.'

'How so?'

'She's like a barnacle, that one. Wouldn't take no for an answer. She got me to do this, helped me with doing the forms and all.'

'What does she say about her parents?'

'She says I should call them.' His face fell. 'But I can't. I can't face Bridie and Joe. Colleen doesn't know it, but I stole money from her mum's wallet last time I was round at theirs, and they must hate . . .'

I made a decision. 'Mick, I saw Bridie recently. She asked me to tell you that she sends her love.'

He looked stunned. 'Really?'

'It sounds like she'd like to see you, Mick. That could be something to think about, as you're both going through this process with your claims. You could help each other, perhaps?' Maybe I shouldn't have said that, but I felt compelled. Just as 'connection is the opposite of addiction', I had also seen how it could be a powerful antidote to trauma. Maybe these two could share some regret and grief and help each other heal.

'I'll think about it,' Mick said slowly. 'I might need to wait a bit. But I'm . . . I dunno . . . We'll see.'

I was aware we'd talked for a long time. The sky was darkening outside, and we had done what we needed to do. I checked to see how Mick was feeling and if he was okay with going home, after all we'd talked about.

He rose, shrugging on his jacket. 'I'm all right. Off to the meeting now, and I'll do another one tomorrow, too. I've got Carmen and everyone waiting for me there. They knew this was on today.'

'I'm glad. I hope this hasn't been too awful for you.'

He flashed his rare grin. 'Better than I thought. That's something, right?' He opened the door. 'Thanks, Mrs.'

'Thank you, Mick.'

———

Both brother and sister were such vivid examples of how childhood adversity shapes and moulds our adult minds. I felt such gratitude that they – and everyone else I met in this process – had dared to come forward and talk about their past. Their stories of identities broken and reshaped would leave a profound mark on me.

Despite his hard path in life, Mick was perhaps more like Bridie than his sister Katie. He had shown resilience, even faith, though he might not have called it that. I thought the IRB would award him a greater level of compensation than his sister because his experience of abuse was long and extensive and his survival had been costly in adulthood, both to him and his family. In my report, I would make a diagnosis of problems with emotional regulation (his depression and chronic anger) and addiction, and confidently assert that there was a causative link between Mick's exposure to childhood adversity and those diagnoses. I knew that his presentation might fit the criteria for Complex PTSD, which was first described by Judith Herman in the late 1980s and captures the accumulation of long-term effects of chronic trauma that leaves people feeling helpless and hopeless.[12] However, twenty years ago, C-PTSD was not yet a formal medical diagnosis, and I thought the IRB might not recognise it as the principal basis of a claim.

My work with Bridie, Mick and many others in that little meeting room above Kilburn High Road confirmed that some healthy aspects of the self can survive even the worst cruelty. The question is, what is the cost of that survival? What is the role of resilience, and how do we measure it?

Nietzsche's idea that we get stronger if difficult experiences don't kill us is often quoted, and that may be true for some people. But I'm not sure Mick would agree. He had lost a great deal, including his childhood. A monetary award based on my report could be helpful, but money might not make a big difference to his life. The definition of 'redress' is to 'make restitution', and I wasn't sure that was possible here; money cannot right wrongs. It was interesting that when she applied, Bridie had a clear idea of how she wanted to use any award she received, but it didn't seem that way for Mick. For him, the biggest gain might be a human one: reconnection with Bridie and her family, and a chance to rebuild some attachments.

As I walked to the Tube that evening, I rounded a corner and found myself outside the local community centre, where a small group of people loitered at the open double doors, smoking and chatting and holding polystyrene cups of coffee or tea. A blue circular sign with an AA logo swung from the door handle. When I paused for a second, a young woman with a knitted cap and a broad smile walked over to me and held out her hand, saying, 'Welcome!'

'Oh, sorry,' I said with embarrassment. 'I was just . . .' I waved and hurried away to catch my train. I hadn't seen Mick among the smokers, but I hoped he was indoors already, in the warmth and the light.

The Hostage

'Nothing happened,' the man was saying, his voice measured and calm. 'Absolutely nothing prompted it.' He smiled at me, as if I might be the one with the problem.

I accepted what he said, smiling back and making a note, but we both knew his GP would not have referred him to me for nothing. This man's denial was intriguing and suggested a degree of avoidance in thinking about his distress.

'Why don't you tell me about it anyway, Douglas? Where were you that day?'

He sighed, then briefly removed his squarish, metal-rimmed glasses, setting them down and then rubbing the bridge of his nose. He was all sharp angles, I thought. Raw-boned and thin, he sat awkwardly in the low armchair facing mine. In combination with a shock of reddish hair and a liberal smattering of freckles, I thought he seemed younger than his fifty-odd years, with the air of an ageing teenager. I waited for him to say more, watching as he took out a handkerchief and polished the lenses of his glasses before replacing them. I was reminded of something I'd heard on TV, in an interview with a young actor. He'd said that fiddling with his glasses helped him settle if his mind went blank mid-scene: 'It makes me look like I'm pausing deliberately, when inside I'm freaking out.' I wondered if the same thing might be happening here.

'I was at the airport. We often go to Frankfurt for work, at least once a quarter. It was totally routine.'

'We?'

'My boss, me and Brian – he's in compliance. We have a routine: we meet a little early, get a full English at a café, go over anything we need to discuss for the meeting and then check in. No big deal. Like I said, nothing unusual. At all.' He was taking such pains to emphasise how illogical it all was that I had to ask myself, Who was he trying to convince? I knew that traumatic memories often defy reason, but I guessed that Douglas might not.

'Your GP mentioned you haven't been on a plane since then,' I said, keeping my tone light, free of anything judgemental.

'That's right,' he mumbled, looking down at his feet.

His GP, Jen, was an old friend of mine from medical school. In the intervening decades, we had regularly kept in touch, meeting in London whenever our busy lives allowed us to. When I'd last seen her, she'd told me that though she knew I did very little private psychotherapy, there was a patient of hers who she thought might need my help. This middle-aged man, in otherwise good health, had experienced an acute panic attack at an airport and was now anxious about flying, which was affecting his work. She felt he ought to see someone sooner rather than later and was concerned about the time it might take for the NHS to respond.

She was not wrong to be worried. The austerity measures introduced in 2010 by the UK's new coalition government had meant massive cuts to mental health services.[1] The financial restrictions, combined with the application of crude health-care business models, meant it was becoming almost impossible for people to access specialist psychological therapies on the NHS. Even the so-called 'talking therapies', which people could refer themselves to without seeing their GP, were oversubscribed, and they could not deal with complex cases. I suspected Douglas's was not straightforward, if only because it is uncommon for people to have a big psychological event out of the blue in their fifties, especially if they have no history of mental disorder. I was working only part-time in the NHS at this stage and had a small private practice that offered extended assessments rather than long-term treatment in order to help people identify the right therapies for them. I agreed to see Douglas on this basis, and we made an appointment to meet in a health and well-being centre, which had rooms that I used for this purpose.

Now, here we were, on a rainy spring afternoon, with me trying to gain some understanding of a difficulty that Douglas was doing his best to minimise. I cast about for something to ask him that wouldn't require a deflection, something easy enough to answer, and realised I didn't know what he did for a living. 'Do you have to fly for your work, Douglas?'

'So far, I've been able to stay in the office, manage things by conference calls . . . Work has been very good.' He paused. 'I've never loved flying, to be honest, but I've done it for years. Part of my job.' He explained that he was a strategic controller for a large City insurance firm. 'We've got offices all over Europe. I'm on audit liaison every quarter. It's never been an issue before.' Again, an emphasis on the ordinary and his professional capability.

I asked if he could take me back to the departure gate, to the moment it happened. Could he remember what he'd been thinking about in those minutes just before he collapsed? Douglas frowned. 'Not really. I was just chatting with Brian, waiting to board.'

'Any physical sensations?'

Douglas thought about this. 'Maybe I felt a little agitated and jittery. Later, I thought maybe I'd had too much caffeine. Could that be a reason?' I didn't answer, waiting for more information. He sat back in his chair and looked away from me, reliving the moment. 'I remember . . . it was like . . . I couldn't breathe. There was this heaviness in my chest . . . and then I thought, Okay, I'm having a heart attack . . . And then there was a huge sense of dread, too awful for words, like nothing I've ever felt before.' He shook his head and continued. 'I was sure I was going to die . . . and then I felt faint. I started to fall . . . I reached out . . . and then . . . I think I must have blacked out for a second. When I came to, Brian and a nice BA woman with a glass of water told me the paramedics were on their way. And then the rest is a bit of a blur . . . The worst was over by the time the ambulance came. They checked me out at A&E, and Brian made sure I got home after.'

His eyes met mine. 'Doctor, even thinking about getting on a plane again scares me shitless. Sorry, I don't mean to—' I waved my hand to indicate a lack of concern about swear words in this context. But I did make a mental note to come back to his 'shit-less' fear because anxiety can often manifest through digestive problems. Maybe there were other signs of physical stress that he hadn't connected to his flying phobia.

'Did you talk about the incident with anyone?'

He nodded. 'I told my wife I'd been taken ill, but a doctor had seen me and said I was fine, and I said the same to work. I felt a bit embarrassed by the attention, the ambulance and all, especially when they said it wasn't a heart attack.'

'What *did* they say it was?'

Now Douglas looked discomfited. 'They called it a panic attack, asking me some of the same things you're asking . . . Had it hap-pened before? Did I have a fear of flying? They wanted to know if I was on any psychiatric medication, too. That made me nervous, being seen as a psychiatric case, you know? No offence.' I shook my head – none taken. Then he added, 'It's just that I've not had much to do with psychiatrists and don't really want to start.'

Yet here you are, I thought to myself. The first psychiatrists were called 'alienists' (from the Latin for 'other'), and mental health services have long been seen as alien worlds that people are not eager to visit. Having a 'mental problem' could be unsettling and might add extra anxiety, at a time when they were probably already upset and disorientated.

'And now? How are you feeling?'

His answer came fast and vehement, and for the first time, his voice sounded loud in the small, sparsely furnished room. 'Fine! I'm fine!' After a beat, he added, 'I haven't had another attack or anything since then, if that's what you mean. This all happened two months ago.' He underlined once more that 'Honestly, I'm fine.' There it was again. I remembered Tom the POW insisting he was fine, too, until he was willing to face his 'feelings inside not expressed'. I knew that took time.

'Really, Doctor, I'm not even sure I need to be here.'

It sounded to me like there was a plea in Douglas's voice, as if he were begging me to release him, but I had to stay with what he'd said. 'Douglas, I think perhaps you would like me to reassure you that you don't need help. But you've also told me that the incident has made you "scared shitless" of flying. Which is a new problem for you?' He shrugged, looking away as though he badly wanted to talk about something else, and I decided to pivot to the question that people couldn't easily gloss over: 'Can you tell me what the worst thing about it was?'

His eyes met mine, windows on his fear. 'The worst thing? It was the sensation that I couldn't do anything. I wanted to, but I just couldn't. In fact, it felt like I *had* to do something. But it was like . . . I was paralysed.' He broke eye contact, hanging his head and studying the floor, his voice dropping to a mumble. 'I mean, I couldn't even tell the paramedic how frightened I felt. I just couldn't. It was awful.' The thought crossed my mind that he might feel shame for some reason, but I set that aside for later.

So far, this was a pretty textbook account of the experience of having a panic attack, which is one of the many bodily manifestations of anxiety. Anxiety can mimic aspects of other diseases or make existing physical symptoms of any kind worse. Studies of anxiety disorders and their treatment mean that all doctors can recognise the physical signs of a panic attack. They are so well known that they have even made it into TV dramas like *The Sopranos*, which opens with mob boss Tony Soprano going to therapy after having an experience similar to Douglas's. I thought the use of the word 'attack' was interesting, conveying not only powerlessness, but also a sense of being victimised by a predator you cannot see or ward off and which means to cause you grave harm.

Now Douglas looked miserable, but I took this as a good sign: he was accepting that something scared him. 'I've got to do something. I can't lose this job. I like the work, the company . . . and we need the money. Maggie, my wife, was made redundant a few months ago, and we've got school fees for my son, you know, and all the rest. I've got to be able to fly again, and my doctor thought you could sort this. Can you?'

'I can certainly think of some therapies that might be helpful, and in my experience, having panic attacks is a problem that responds well to treatment. But we need to understand more about why this has happened to you now. Tell me, is there something in your history that's relevant here? Something that would make you afraid of getting on a plane?'

Douglas sounded slightly irritated. 'I mean . . . yes. Something happened to me on a plane once. But it was aeons ago; I was a boy. And I told you, I've been flying for years. Why would it suddenly become a problem now? Makes no sense.' Interesting – I suspected he'd not told Jen about this 'something', or else she would surely have mentioned it. And still there was that emphasis on the irrationality of his experience. How much easier it would be if we did not live in a world that constantly tells us that logical things are the norm and somehow 'right', while chaos or surprise are abnormal or 'wrong'.

We were coming to the end of our first meeting, and I wanted Douglas to leave with a plan. I suggested we meet again, and in the meantime, I would contact a colleague of mine who offered cognitive behavioural therapy (CBT) for panic attacks. I knew Douglas would google it as soon as he got home, so I simply explained that CBT was a therapy first developed in the 1970s to help people better understand their thoughts and how they influence emotions and behaviours.[2] CBT is a useful intervention for anxiety states because it allows people to gain a sense of control over what is going on in their minds, which reduces distress. They learn to recognise the unhelpful beliefs and distorted thought patterns that can activate crippling fear or sorrow and then begin to challenge them.

By the time I met with Douglas, there was good evidence that CBT was an effective mental health intervention that helps people feel less anxious and improves their mood. Today, it is pretty much the only therapy freely available on the NHS because it's short-term and comparatively easy to train therapists to deliver it. Many people need only a few sessions to see results. When he heard this, Douglas seemed open and eager to try it. Then I added, 'But if you are willing, I'd like to hear more about your past experience on a plane. I understand it was a long time ago, and maybe it's not been on your mind for a while, but from my perspective, it would be worth exploring and could be helpful for your CBT therapist, too. Are you okay with coming back to see me to do that?'

Douglas looked mulish. 'I don't know if I can remember much about it – it was so long ago. I just want to get on with this . . . I don't want any long-term therapy or drugs, you know?' I noted his resistance, gently commenting that I wasn't suggesting either. He said he was glad of that and even smiled when I gave him an appointment card, tucking it into his wallet as if this were proof that he would 'get sorted' (I've seen prescription slips provide similar comfort). He told me he'd see me the following week and thanked me for my time.

But he cancelled our next appointment, and we didn't meet up for nearly a month. During that time, I spoke to Philippa, a psychologist colleague who was also working in the health-care clinic where I had rooms. I knew she offered CBT for different phobias and anxiety, including panic disorder, and she said she would be happy to see Douglas, if he got in touch.

The next time Douglas and I met, I could tell from the moment we began that he was uncomfortable. I decided to name this problem from the start: 'Douglas, I am guessing this process is hard for you, because it's been a while since our first meeting. It's brave of you to come back and talk with me. We can take this at your own pace. Can you tell me about what happened when you were a boy?'

He cleared his throat, then dived in. 'Okay. As I said, I don't remember too much. We'd been to see my dad, who was working abroad. I was in boarding school in Hampshire then – I would have been thirteen, nearly fourteen. Mum was taking me back for the start of term, when the plane we were on was hijacked.' I was startled and said something to the effect that I had never met anyone who had been through such an experience before. Douglas looked down, seeming almost irritated by my interest. 'Nothing happened, though.' 'Nothing happened?' I echoed. 'Well, it was only a few hours – they didn't harm us. They let the women and children go after a bit. Then I guess they made a deal, and everyone was freed. We flew home. That was that.'

Douglas's tendency to downplay his own distress or fear was familiar to me now; it was part of his way of coping. It was important not to undermine him at this stage, but to try and make him feel safe enough to be more open. I was glad I'd set aside some extra time for this session. 'That's a helpful summary, Douglas, thank you. What else do you remember?'

He sat up straighter, smoothing his shirtfront, then running a hand through the sweep of hair that fell across his brow. 'I remember . . . let's see . . . The whole process of flying was quite familiar because Mum and I had done it several times by then. Of course, it was simpler back in the day. I'm not even sure they had metal detectors or anything. We just stood in line and walked out on the tarmac to the plane, right?' I nodded. I'd had a similar experience as a child.

Over the next hour or so, Douglas relaxed a little and moved on from making general observations about flying to his hijack story. It was clear that despite his previous assertions, he did have remarkably detailed memories of that day. 'It's funny how certain things stay with you, isn't it? Like, I remember the flight took off precisely on time. I'm sure of that because I was obsessed with my new watch . . . "The new, self-winding Seiko Sportsmatic!"' He was consciously speaking like the boy he had been, and I smiled with him. 'I still have it somewhere. Thought I'd give it to my son, but he doesn't . . . well, kids, you know? They want the latest thing.'

'And then?'

He sidestepped, moving away from the personal. 'Air travel was so different then, wasn't it?' He spoke about the sense of occasion it once had, the formality of all the male passengers wearing suits and the women in smart dresses and even hats, their children on best behaviour, kitted out in their Sunday best. 'We were more like mini-adults, then, I'd say.' I thought I knew what he meant but waited for him to elaborate. 'You know, please and thank you, sit up straight, speak when spoken to, no fuss. Nothing like now. You should hear my son—' He broke off, frowning. 'I'll get to him later.' I noted the tone of his voice and the slight sense of threat about what we might 'get to' later. He talked about a willowy, smiling flight attendant, too, who stood greeting people at the door of the plane as they boarded. 'I thought she was a dead ringer for Ursula Andress,' he smiled, a reference to the most famous 'Bond girl' of our youth. He stopped there, as if reluctant to take me through the door and on board the ill-fated plane.

'Did you fly often as a child?' I asked. He seemed relieved to divert again and spent some time talking about his experience of boarding school, which he'd attended since the age of eight, after his father, a geologist, got a job based in the Middle East. When I asked, he described it as a good experience overall. His mum lived not far away, and he saw her regularly. They always spent the holidays together in England or with his father abroad. Although not for all children, boarding school has its positives, including a secure base and lifelong friends. But it can mean that pupils learn to manage distress on their own, which might also explain some of Douglas's reluctance to seek help from me.

I encouraged him to return to his memories of the flight, and he seemed ready to do so, although he stuck with the easy part of the story, setting the scene rather than skipping forward to the main event. Their seats were near the front of the plane, he told me, not far from the exit doors. His mother was given a glass of champagne, and Douglas was offered a chance to go into the cockpit and meet the pilot. The co-pilot introduced himself as 'Captain Mike', and he let Douglas wear his uniform cap for a minute. A couple of other young children, doubtless also heading home after the holidays, came in after him, and they were all given metal wing badges to pin on their blazers. As Douglas described this, I also remembered getting one of those badges as a child, proclaiming me a proud member of the Junior Flyers' Club.

Another detail that Douglas remembered was that in the row behind them, a young mother was tending to a fractious toddler who was wiggling and fretting as everyone settled down for the safety demonstration and the buckling of seat belts. He also recalled that two men in dark suits were sitting in the row on the other side of the aisle, and one of them leaned across to get his attention, touching his arm. 'He had a thick accent and pointed at my new watch, asking, "Time?" I showed the watch face to him, and he said, "Good," and thanked me, then turned and said something to his seatmate, speaking in a language I didn't recognise. I thought he was telling him he liked my watch.' Douglas shook his head. 'I was so pleased with that thing.'

Soon, the almighty roar of the engines drowned out all further speech. It was a long journey with multiple stops; the next one would be Delhi. Douglas remembered how his mother had tried to sleep, while he listened to comedy shows on the in-flight audio system; that was his introduction to the Goons, Monty Python and *Round the Horne*. He wasn't sure how many hours had passed when the man across the aisle tapped him on the arm again, smiling and nodding. He was holding up the in-flight magazine, which was turned to a page with a map, and seemed to be asking if Douglas knew where they were on the flight path marked with a red dotted line. The request made him feel important, a well-travelled young man. His fellow passenger also wanted to know the time again, and Douglas obliged. 'I remember feeling a bit superior . . . even a bit sorry for him that he was so clueless. Seriously.' Listening to this, I felt apprehension, even dread, as I waited for his world to change from the pleasantly familiar to something disastrously new.

Out tumbled more details: how Douglas hoped to see the pretty air hostess again and how he had to push through a curtain of cigarette smoke to get to the loo at the back of the plane. 'I remember I almost got stuck in the little compartment. I couldn't open the stupid door for a minute or two. I was terrified I might have to call for help.' But a different terror was about to begin. 'When I finally got out, I realised that something had happened . . . Everything had gone very quiet, nobody was chatting. And everyone was in their seats, with their hands on their heads. I remember some of the women and children were crying.'

He was confused more than scared when the man he'd engaged with across the aisle strode towards him, grabbing him roughly by the arm and yanking him forward. Then he saw that there were now four or five men on their feet, all armed, with guns pointed at the passengers. Douglas was unceremoniously deposited back into the seat next to his mother. Her hands were atop her head, too, her eyes squeezed tightly shut. 'Mum?' 'Shh, Dougie . . .' she said. 'Shh. Hands on your head.' Slowly, Douglas complied, just as the toddler behind him began to whimper and complain, kicking at the seatback.

'I remember one of the men barking "Silence!" at the child's mother, pointing his gun at her . . . I remember her repeating, under her breath, "Come now, I'm begging you, I'm begging you" to the child, over and over . . .' Douglas paused and swallowed. I was glad I'd put out a glass of water for him, which he grabbed and drank from before he resumed his story.

'Then I saw the man who had asked me the time, standing behind the air hostess . . . He had her in a chokehold, a knife in his hand. She looked terrified . . . so helpless. And I remember the captain's voice speaking to us all over the intercom, telling us that we were going to make an emergency landing and that we needed to do as we were told.'

Douglas took another sip of water. 'I think that was when I realised we were in mortal danger . . . I had never seen my mum scared before, you know? And everyone's still sitting there with their hands on their heads. It sounds easy, but your arms start aching before long, and you don't dare rest them in case they . . .' He looked away then, unable to continue, and I sat quietly, letting him take whatever time he needed to manage his recollected fear.

I thought of all the hours I'd spent on planes in my childhood, often bored and lonely and anxious. Such forced inactivity can be stressful, even without the extreme complication of a hijacking. I was imagining how hostages are frozen during their ordeal, suspended like insects in amber, with no idea of the endpoint or outcome. Most ordinary prisoners at least have some idea of how long they will be held in forced confinement; the hostages Douglas described had more in common with wrongly convicted prisoners or people detained in immigration centres, who speak of how their confinement feels infinite.[3]

Douglas had fallen silent, his gaze directed past my shoulder and out of the window, towards the London sky. After a minute or so, I prompted him to go on: 'And then?'

He seemed to shake himself back into the moment. 'What? Oh – well, eventually, we landed. Once we were on the ground, on some airstrip in the middle of nowhere, the hijackers started moving people around. All the men were ordered to go to the back of the plane, women and children to the front.'

Since he didn't consider himself in the latter category, Douglas had jumped to his feet. 'Sit down! Men only!' Someone pushed him back down. It was the same man who'd been threatening the air hostess, the man he'd helped with the map – which made him feel outraged; 'conned' was the word he used.

'Conned?'

'He was pretending to be friendly, wasn't he? And all the time, he and the others were planning this thing . . . He'd talked to me like a man, and now he was treating me like a kid. "Men only!"' Douglas stopped, his face flushed. 'Everyone heard him say that to me, including my mum. I remember she was cross and hissed at me, "What were you thinking?" I was mortified . . . and completely helpless.'

Other distressing memories came back quickly now. Douglas remembered a couple of the male passengers protesting at being separated from their wives and families. An older Scottish man of military bearing, who reminded Douglas of his father, demanded to know what the devil was going on, as if they were on a delayed train or the waiter had forgotten their order. One of the hijackers promptly smashed him in the face with the butt of a rifle, and Douglas remembered seeing the blood pouring from his nose.

Then he noticed two of the hijackers were busying themselves with some rope. He was holding his breath as he watched them attach two black cylinders – 'like Coke cans' – to it, before looping it across the emergency exit door. Were those bombs? They were right at his eye level, and he could see wires and some sort of timer. 'Mum? Are we going to die?' Douglas recalled whispering. 'And I remember what she said so clearly. She shut her eyes again, leaned back in her seat and whispered, "I don't know."'

This was a universal terror, the kind associated with maximal distress. Douglas said he closed his eyes, too, mimicking his mother, as children do in order to learn how to navigate the world. It was quite an image: mother and son side by side, frozen in the face of their shared fear, trying not to see the end of their lives.

It now made absolute sense to me why the adult Douglas thought he might die if he got on a plane, but there was still much more to hear. He recalled the long – who knows how long? – period when they sat on the tarmac in the sweltering, unfiltered air of the stationary plane, not knowing what was going on. Suddenly, Douglas said something I wasn't expecting: 'I felt so guilty.'

'About?'

'That man . . . he asked me the time, and I told him. I'd helped him take control of the plane. All those people . . . I felt like it was all my fault.'

He was tearful then, just for a moment, before he stopped himself, and I remembered how he told me he had tried not to cry on the plane, in case that drew more attention to him.

'What else is coming to your mind?

'There was a smell of urine . . . and that baby was still crying and fretting . . . Her mother was getting more and more upset, shushing and pleading . . . Later on, I remember seeing the welts where the woman dug her nails into the child's arms to try and silence her.' Desperation made visible.

It all sounded unbearable – the sensory overload of smells, heat and sound, the fear, the lack of information, everything. And then there were noises outside, the shuddering sound of rolling portable steps on tarmac, and the forward door of the plane opened. Captain Mike's voice came over the intercom, instructing all women and children to exit the aircraft, leaving their belongings and walking in single file. 'Please continue to keep your hands on your heads, everyone.' In silence, they followed his instructions, some of the women swivelling around for a last anguished look back at their husbands, colleagues, brothers or sons, clustered at the back of the plane, hijackers standing over them with rifles drawn. Outside, in the blast of humid air and brilliant white sunlight, Douglas saw another plane, waiting for those who had been released. 'We were all flown to a military base in Germany. I remember that no one talked on the flight, as if we were still hostages. I felt like I was holding my breath the whole time.'

Again, this recalled his recent experience at the airport as an adult. But I was puzzling about why this had never happened to him before. He was describing something that would leave many people with enduring post-traumatic stress symptoms, including panic attacks and phobias about anything that might be a reminder. He had been stuck for hours on a hijacked plane, hands on head, arms aching, terrified that death was imminent, unable to do anything. He described feelings of fear, shame and guilt, but by his account, he had never spoken of them before. Where do such feelings go if they aren't articulated? How long will the dam hold? Sometimes people who self-harm describe a similar sense of being stuck in helpless terror. I've had patients tell me that they feel they must cut themselves or pierce their skin, as if the open wound will allow bad feelings to escape the body like steam from a kettle. What happens if you can't get that relief?

Douglas's next memory was of speaking to his father on the phone at the German airbase. The crackling voice coming down the line was calm but unmistakably relieved. 'Jessie? Dougie? Are you all right? Thank God. Thank God you're safe.' Douglas remembered that he started crying, grabbed the handset and sobbed, '"Daddy! Daddy!" And I remember that he said something like, "Don't cry, son, there's a good lad. Put your mother back on." Mum took the phone back, saying, "He's doing fine. We're both fine. Why wouldn't we be? We're safe now."'

I thought that might be a good place to stop for the day – landing in a place of safety. Douglas agreed, but when I commented that he had worked hard and looked a bit tired, he dismissed this thought politely, saying, as his mother had once done, that he was 'doing fine'. He told me that he had arranged an appointment with Philippa for their CBT work to begin soon, and I checked whether it was okay for me to share some of this story with her in advance so that he needn't go over it again. Douglas was happy for me to do so. I warned him that some memories of his experience might return over the next few days, but he seemed unconcerned and looked calm and composed as he thanked me and departed.

He'd given me a lot to think about. A case like this was new to me. The only 'hostages' I'd seen had been women and children exposed to repeated domestic violence, but their context was very different; the perpetrators were husbands and fathers, not complete strangers. At one of the trauma clinics, I had worked for some time with a man who'd been held as a prisoner for years in an Iraqi jail. Nabil's family had been politically active opponents of Saddam Hussein's regime, so his imprisonment – at the tender age of eighteen – was intended to put pressure on his family. He was held in a tiny underground cell with many other men and tortured and beaten on a regular basis, until his family paid a huge ransom for his release. He and his relatives then moved to England, where he was referred to the clinic as part of its service to refugees.

You could say that Nabil had PTSD, but that wouldn't capture the full extent of his suffering, of how broken he seemed. His chronic terror had a paranoid intensity, and regular outbursts of rage got him into daily trouble with other people. Although he could speak English reasonably well, our work was challenging due to the incoherence of his mind, as if some capacity for speech had been shattered. But we spent many hours sitting together, with me just listening and trying to make sense of his evident fear and shame. He was one of the saddest men I've ever met.

He and Douglas had something in common: they had been bargaining chips in someone else's political game, impotent in the face of events around them. Both had been lucky to survive their ordeals, but the former Iraqi prisoner had not been able to put his horrific experiences in the past. In marked contrast, Douglas was articulate and high-functioning; it was as if his trauma had happened to a different person. I reflected that this might be because his experience had been relatively brief; it could also be that he was protected somewhat by having gone through it with his mother at his side, with whom he seemed to have a secure attachment. That may have given him some resilience to call upon, although even as I had this thought, I remembered Lisa, the helicopter crash survivor, who had been 'fine', too – until she was not.

I didn't hear from Douglas for a while and assumed all was progressing well with his therapy. Then, out of the blue, Philippa left a message on my phone, asking if she could 'have a word'. Something about her voice made me return her call immediately – she sounded uneasy, shaken. The moment I said hello, she blurted, 'Gwen, there's a problem.' Douglas was *definitely* not ready for CBT.

'Why is that?' I said, a bit puzzled. 'He seemed keen to try it, especially as he's got to be able to fly for his work.' Philippa explained that the first couple of meetings with him had been productive; she'd explored some of the same issues he'd discussed with me, and they had planned a series of treatment sessions. She had given him some monitoring homework in the form of a diary, in which he could keep a note of his moods and thoughts, and he had done well with this. But she noticed he was increasingly irritable each time he came in and struggled to keep his temper. In their most recent session, the day before, while she was trying to work through with him how his thoughts and feelings were linked to an issue with his son, he had become agitated. He'd risen to his feet, pacing up and down, and she hadn't been able to soothe him.

'Did he raise his voice?' It's comparatively unusual for people to do that with therapists.

'He absolutely did, ranting, telling me I didn't know what he was going through, that I was clueless. I kept asking him to sit down so we could talk, but that made him even angrier, and then he swore at me, punched the wall and slammed out of the room.' Ouch. I imagined his knuckles might be worse off than the wall. I shook my head; this was not at all expected. At the same time, it occurred to me that by acting out with his therapist in this way, he was behaving rather like a terrorist, creating sudden panic and threat in a contained space. 'Honestly, Gwen,' Philippa said, 'I don't think I can see him again. He might be too risky, if he can't regulate his anger. Does he need some medication to get his mood stabilised before he attempts any more therapy?'

His outburst had frightened her. I knew she had not worked much with men before; her training placements had been in services where the patients were mainly female. That's not unusual for trainee psychiatrists and psychologists: the data shows a marked gender disparity in people accessing mental health services in the UK; one recent survey indicated that women seek help twice as much as men.[4] Philippa went on to say that when Douglas punched the wall, it occurred to her that he might hit her, too. 'I've never been frightened of a patient before . . . Have you?'

Given that she knew I'd worked at Broadmoor, I imagine she thought I'd tell her about some incidents of physical violence or threat, either there or in a prison. But the truth is, these secure places are generally safer to work in, as everyone is prepared for risky behaviour, and we anticipate it if we can. It is in community clinics and public places where violence can erupt out of nowhere. I did share a vivid memory with her from my trainee days, of a woman in a highly distressed state of mind running rampant through an outpatient clinic while waving a razor blade. I had bravely hidden myself in a stationery cupboard, until the police came and took the woman away. I wanted Philippa to know that I understood what it was to feel frightened by another person's unexpected anger and ill-equipped to respond. I would follow up with Douglas myself, I told her. This development suggested he was struggling not so much with a phobia or panic issue as a problem with unresolved anger.

———

I wasn't sure if Douglas would return to see me, if invited; the interaction with Philippa might have left him agitated and ashamed. I felt pleased when he responded and came back to meet me a few weeks later, full of apologies and deeply embarrassed. Right away, I told him I thought I had missed something in the original assessment that I should have picked up on; perhaps I had focused only on his fear and distress, without exploring the anger and rage he might have felt. Douglas had shown Philippa that he had the capacity to be both angry and frightening, and that was where we needed to start. I thought, but didn't say, that shame often manifests as rage.

As we went over the incident, I remembered his description of the beautiful flight attendant staring at him in terror from the front of the plane, a knife at her throat. I tried to get him to think about what he'd felt about Philippa when he punched the wall in her office. Douglas looked stricken and shook his head, struggling to find words. All he could say was that the CBT work was pointless; he was starting to think he would never get better and felt he had to do something. He remembered getting up from his chair and feeling furious when Philippa told him to sit down. He hadn't wanted to hit her, but he had to express this rage somehow, and the wall was there. And then he had to escape.

I began to think that what I'd missed in his self-narrative was Douglas's need to always be in control, especially if he felt vulnerable, and that took us back to the hijack. I suggested we pick up where we'd left off at our last meeting. 'You were telling me, Douglas, that once you were safe in Germany, you and your mother had rung your dad, and you were crying . . . What happened then?'

Douglas thought for a moment. 'Just then? She took the phone away and shooed me off.' Then he recalled something else. 'Oh – and I could hear her telling my dad how I'd "made friends" with the hijackers when we boarded the plane, how I'd been "merrily chatting away to them" before the hijack . . . as if that was true . . . kind of like it was a joke. I didn't understand how she could do that after the awful experience we'd just had. She was acting like I'd done something funny . . . almost like she was mocking me.'

This was interesting; that could have left him feeling hypersensitive to being misunderstood by people he trusted when he felt vulnerable, particularly women. Did that have some connection to what happened with Philippa? I didn't put this to Douglas, but I did ask if he thought his mother's fear of a violent death might have made her (at least temporarily) unable to care for him in the way that he needed. He shook his head and said he didn't know; he'd never looked at it like that. Was it possible that in the aftermath of the hijacking, his mother had been trying (albeit clunkily) to lighten the mood and perhaps reassure him, his father and herself that it was all over and had not been particularly serious? And

I added, 'As parents, we don't always get it right for our children, I think.' He smiled ruefully, and I continued. 'Douglas, did you ever tell your mother how guilty you felt about talking to the hijacker, giving him the time and showing him where you were on the map? Maybe when you were safely back home in England?'

'Not really. I couldn't stand for her to go on about it – or refer to that man as my "friend" again. My dad mentioned that, too, the next time we spoke, teasing me about it, which made me squirm.'

'Because?'

'All I could think of was how I had let people down by helping those men, that they had conned me and then treated me like a child. I wanted to be brave, but I wasn't, I was . . .' – he paused to find the right words – '. . . just a helpless little boy, with no one to protect me.' It occurred to me then that he may have felt anger towards both of his parents for not protecting him, even though he knew that wasn't rational. Your life depending on someone who frightens you arouses complex emotions and a wish to please – another reason why jokes about 'your friend the hijacker' could have felt hugely shaming.

I thought there was something here, too, about ideas of masculinity and strength. Douglas had likened the flight attendant to an actress in a Bond film, and I thought about how Sean Connery's 007 (Scottish and ex-military, like Douglas's father) always met *his* life-and-death challenges with great aplomb and wry humour. Had the hijackers embodied an all-powerful masculinity that he lacked and an imperviousness to danger that his parents didn't have either?

'What happened next?'

He leaned back in his chair. 'Um . . . Mum and I flew on to England, and I went back to school. She must have told them what happened, but I started the spring term as usual. I didn't talk about it with anyone there. I mean, I couldn't say a word. To begin with.'

'What do you mean?'

'Just that . . . I couldn't speak. I lost my voice. Not like laryngitis or anything. I became mute for a while.'

Now I was intrigued. This phenomenon – post-traumatic aphonia – was well known in the literature, although I had not seen it before. I had learned from Pat Barker's books about the First World War[5] that it had been common in soldiers with what was then called 'war neurosis'. Aphonia is quite different from the emotional speechlessness that former patients like Nadia and Evan had described to me; it is more of a physical reflex in the immediate aftermath than a chronic issue. In another case, I'd seen someone who had temporarily lost the use of his legs in response to extreme fear, although he fully regained his ability to walk after a week or two. Was it like that? Douglas confirmed it was: his voice had vanished immediately after returning to school, then came back just as unpredictably within a couple of weeks.

I thought about whether he had unconsciously muted himself. Obeying orders to be silent had kept him safe while he and his mother were hostages. Douglas's parents had also unwittingly silenced him by making light of his distress and joking about his 'friendship' with the hijackers, thereby offering him neither comfort nor the safe expression of ordinary fear. Silence can be a refuge when any communication feels impossible.

Douglas's experience of aphonia also made me think about silence as a form of control. The hijacking happened just as he was going through puberty, a time when young people often feel out of control, either physiologically or psychologically. Sitting on that plane, testosterone and adrenaline surging, Douglas was both a young boy who needed protection and a young adult who was longing to be seen as a man who is in control. In that state, he may have been especially susceptible to the demonstrations of dominance exhibited by the men who took over the plane. He both did and did not want to be their ally. He did not want to be their prisoner, he wanted to be in charge, as they were.

'And you didn't talk about the hijack?' He shook his head. 'What about when you went home at the end of term? Did it come up then?'

'Not really, beyond details about Dad's discussions with the airline about compensation and such . . . I think we kind of agreed that it happened, that it was unpleasant, but it was over and done with. I just wanted to forget it and get back to being a normal kid again. It all got packed away in the past.'

'Packed away': as if trauma were something we bundle into a chest of drawers or tuck into a box in the closet. Evan, the son of the Holocaust survivors who hid their documents and letters in the attic, had used similar language.

'And then?'

'And then Dad got a new job back in the UK, and we all moved to Scotland. I had to change school, which was all right by me.' He stopped short. 'Now that I say it, maybe he put in for the transfer because of the hijacking? I've never asked.'

'And did you ever lose your voice like that again?'

He shook his head vehemently. 'No. And soon enough, I was fine.' That placeholder word again. He liked his new school in Scotland, enjoying sports and excelling in maths and music, eventually opting to attend university at St Andrews. He studied accountancy and found a job with a prestigious firm in London. There, he met his wife, Maggie, and they had two children: a girl, Ailsa, and a boy, Douglas Jr (known as Dougie). He saw his parents occasionally; they still lived in Scotland. His mum was in good health, while his father was increasingly frail. 'I've always flown up to visit them, by the way, and never had a problem.'

We left it there, and he agreed to come back the following week.

———

Later, I reflected how Douglas's story demonstrated that trauma is not a discrete temporal event. Equally important is what brackets it: who and where you are as it starts and what happens afterwards. For him, the after-time was significant because the silencing he'd experienced on the plane did not end: his voice was quashed by his parents and by his body with the aphonia. I saw those events as a continuation of the trauma on the plane. Only when he was living back at school, apart from the family, could he react to the totality of what had occurred, and then he was struck silent.

In our next session, I wanted to explore whether the hijack experience had changed his parents' relationship or anything else at home. He had to think about that for a bit. 'Well, before the hijacking, they never rowed; as far back as I could remember, they were always calm and polite with each other, at least in my pres-ence. Good manners, good breeding, you know. They both came from pretty uptight backgrounds: my grandfather on my mother's side was a minister in Edinburgh, strict Free Presbyterian; and Dad's from an army family, generations of tight-lipped officers. As a child I don't think I ever heard Mum raise her voice, although Dad had a bit of a temper. Just not with her.'

I didn't comment, hoping he would say more; was he suggesting his father would turn that temper on him? Interesting phrasing, too: what does 'a bit of a temper' mean? What would a whole temper look like? 'Anyway,' Douglas went on, 'after he came home that summer, I noticed they had a few rows, once we settled into our new home in Scotland. Nothing huge; stupid things, but I'd never seen her snap at him or criticise him before. It seemed like she was always on edge; the slightest thing would set her off.'

That metaphor of uncontrolled anger naturally recalled those excruciating hours on the plane. So many media narratives of trauma survivors portray them as cowering, helpless victims, generally female, moving anxiously through the world, haunted and hushed. But I reminded myself that anger is the primary response to fear or pain. I thought Douglas was describing a woman who was in a high state of irritability and perhaps hypersensitivity to threat. Douglas had similarly expressed his anger in the room with Philippa. I was curious as to whether I might see it in our work together, too. I didn't have to wait long.

————

He came in looking a little dishevelled that day and immediately announced he was feeling 'knackered'; as so many trauma survivors report, he had not slept well. 'What are we talking about today?' He was restless in his chair, changing position a few times, taking out a handkerchief, then pocketing it again. 'What's on your mind, Douglas? Is something in particular bothering you?'

He made a peevish gesture, swiping the air, saying he had no idea.

'What's going on?' I tried.

'I'm fine,' he retorted, his voice thick with sarcasm. 'Just a little tired. Can't you see?'

If I were seeing him now, I think I would try to get a handle on what was happening by using language to describe my experience of him, along the lines of, 'I'm getting a strong sense that something I'm saying or doing is irritating you.' In the moment, however, I felt anxious not to upset him and make things worse, and I think he picked up on that. He stood abruptly, moving to the window, his arms crossed over his chest. 'The thing is, I think I need to go to another therapist, someone more senior, someone with proper experience. This is just like with that other young woman. She didn't know what she was doing either.' His voice rose like a command, some menace creeping in. 'I want you to refer me to someone else, because these sessions are useless. Pathetic.'

There were so many irrational messages here: that I was inexperienced; that I was a failure because I was a woman who couldn't help him; that I was incapable and powerless, and pathetic in my powerlessness. I struggled not to feel belittled and humiliated and sensed my anger rising. But some part of me knew this interaction was a re-enactment of the hijack experience: now, Douglas was the one in command who made other people feel foolish and scared. I had experienced this imposition of status once or twice before with angry men, and I knew that any threat display of my own would only make things worse. I was icy quiet in my response as I worked out how to re-establish a dialogue between therapist and patient, not hostage and terrorist. 'Douglas, I feel uncomfortable when I hear a raised voice, and I can't hear or understand you properly if you stand with your back to me.'

He wheeled around and looked me full in the face. I held eye contact with him and said, again in a quiet voice, using his name to remind him that we were in a personal dialogue, 'Douglas, I think if I ask you to sit down, it reminds you of the hijack and makes you feel threatened. So I won't do that. But getting angry with your therapist will probably not help you, either. What would you like to do?'

Douglas took a couple of deep breaths and remained standing. Time seemed to crawl, although it could only have been a few seconds before he dropped back into his chair. When he spoke, his tone conveyed that he was still irate. 'I don't like being made a fool of or wasting my time.'

I paused, before I responded, 'I hear what you say, and I'm sorry if you feel that's happening right now. I sense you are struggling with memories you once told me you had *packed away*.' I hoped he would remember that phrasing. 'I wonder if our sessions sometimes leave you feeling vulnerable, much as you felt vulnerable on the plane. Pulling out memories of past times when you felt powerless and mistreated might be upsetting in a way that you didn't expect when you came here.'

I had no idea how he would take this, and I remained very still, trying to keep my demeanour neutral while communicating a sense of interest and warmth. I noticed that my anger had melted away. I wanted him to take a risk and become willing to explore what had just happened. To my relief, he was.

'You're right, you know.'

'In what way?'

'There's something about coming here and talking that I find . . . demeaning. I feel tense when I sit down and you bring up the hijack. I feel like I don't know where the session will go – as if I'm skiing downhill with no poles. I know you're not trying to make me feel bad, but . . .'

'Is it as if I have all the control?'

'Something like that . . . I know you aren't telling me what to do, but I feel helpless – or something. I don't know how else to describe it.'

'And it's okay not to know, Douglas.' Once, at a crucial moment, I was offered a lifeline by a wise man, a mentor of mine, and it felt fitting to share it with Douglas now. 'Maybe not knowing is an important step to understanding something in a new way?' I waited for that idea to settle, then added, 'I'm noticing that your anger seems very close to the surface today. Do you have an idea why?'

'There's been a bit of an issue at home again.' I raised my eyebrows, immediately interested. Douglas had said little about his wife and children, except for those few comments about his son at our first meeting. The phrase 'a bit of an issue' reminded me of his recent description of his father as having 'a bit of a temper'. He and his wife had argued at the weekend. On the face of it, the problem was related to their son Dougie, now aged thirteen, who had wanted to go to a classmate's birthday party. Douglas had disapproved; it was at the other end of London, and they didn't know the parents. He had put his foot down. But Maggie had already given her permission for the boy to go, and they started arguing. 'It was just like that bloody school trip,' he said.

'I'm sorry, the school trip?' He'd lost me.

'Didn't I tell you about that? It was a few months ago – just before that thing at the airport. It was quite a row. I mean, they happen pretty often these days. Dougie's at that age, always pushing the boundaries, you know . . . although I wasn't like that at all, I must tell you.' An intriguing comment, but if the fight about Dougie directly preceded the panic attack at the airport, I needed to stay with that. 'Go on. Tell me about the school trip.'

A few months earlier, Dougie had been chosen by his school to go to the Netherlands to play in a football tournament, and Maggie had signed the form that gave parental permission. Douglas had come home from work after a hard day, feeling tired and stressed. When he went upstairs to change, he found Maggie in the study, looking for Dougie's passport. '*Voilà!*' She handed it to him, opened on the photo page, and commented, 'He's the image of you at that age, isn't he?'

Douglas said that when he looked at the passport picture, he felt a rush of emotion – a 'hot feeling', as he put it, a mix of terror, rage and bewilderment. His adult self knew that he was safe at home, but he was flooded with memories of his boyhood self: his enthusiasm for school, his passion for sports, and the prized Seiko Sportsmatic watch he wore every day. He said that he suddenly had a memory of the hijackers forcing everyone to hand over their passports at gunpoint, and how frightening that felt, giving up that emblem of identity and freedom of movement.

His face reddening, his hands shaking, Douglas had thrown his son's passport on the floor and told his wife that he forbade the boy to go on the school trip. She was mystified – 'Come on, don't be unreasonable.' Douglas had snapped at her that his money was paying for it, and he'd decide, not her. He had shouted something like, 'What I say, goes!'

His son had rushed in to plead with him to change his mind, crying and whining, but that only made Douglas angrier. 'Shut up!' he'd snapped, and Maggie had gasped – 'We have a rule that we don't say that in our house,' he explained to me. As the emotional temperature in the room rose and he saw the shocked faces of Maggie and Dougie, Douglas felt the urge to hit something or someone, fury rising like bile in his throat.

He paused, and I asked, in a mild tone, 'And did you hit anything? Or anyone?' His face flushed, and he looked tearful. 'No! I've never hit anyone in my life. I wouldn't harm my wife or my children. I can't bear the thought of it.'

'And the thought was there.' Again, I kept my tone neutral, without judgement.

Douglas nodded, miserable. 'I was shocked . . . but thankfully, the feeling subsided pretty quickly, and then all I wanted was to get away.' He'd stormed out of the house, slamming the door so hard it shook, then spent the night at a friend's, only coming home on Sunday night to sleep in the spare room. 'Not that I slept. I couldn't.' He felt drained and ashamed as he left the house on Monday morning, before the family were awake, going straight to the airport, where he later had the panic attack that led to his referral to me.

'Can you say more about what was going on when you were alone on Sunday night in the spare room?'

In a quiet voice, he told me, 'I was afraid.'

'What about?'

He described a desperation, a terror that he would not be able to control his temper in front of his family. 'I realised that I could have hurt them. I could see the fear in their faces . . . looking at me like I was a . . .' – he hung his head – '. . . as if I was a threat to them . . . which I guess I was.'

I thought I had a metaphor to offer that might help him articulate his difficulty. 'It sounds like you felt something explosive inside you. Like the explosives you saw on the exit door on the plane?' Douglas looked startled. 'Maybe so, yes. I knew I needed to stay in control of the situation; I didn't want to explode . . . I was so worried that I might hurt the people I love.' At this, he began to cry. 'I love my son . . . we get on so well. No father could ask for a better son. We're much closer than I ever was with my father.'

I had a question: 'Were you afraid of your dad when you were little, Douglas?'

He sniffed and reached for a box of tissues on the table, pulling out one, then another. 'He was a bit scary sometimes. I know he loved me, but he was always so far away. Distant.' Perhaps both emotionally and physically, I thought, but did not say, not wishing to interrupt. 'He was always so impressive, really an imposing man back then. In charge. I wanted to be like him, but I was just . . .' He looked up at me, his face grave and sad. 'I was a little boy on that plane that day. I couldn't do anything when those guys took over. They were the ones with all the power . . . and I wanted to scream at them and lash out at them, beat them up or something, but I couldn't. I just had to sit there.'

I let him reflect on this as he dabbed briefly at his eyes with a tissue, then asked, 'Do you think that's why Philippa or me asking you to sit down makes you angry?'

He admitted this was possible. 'It's hard to accept when things aren't in my control. My son's growing up, demanding his independence, wanting to go out in the world. My daughter's off to uni next year . . . I'll not be there to protect them, and I don't like it.'

I had an idea: 'Is it possible your original panic attack was your mind trying to protect you from going into that space where you feel powerless over your anger? As if you might explode? And then this weekend, that feeling came up again, with Dougie pushing the boundaries, and . . .'

Douglas finished the sentence: '. . . I wanted to push back. Push him down.' I flashed on that memory of the hijacker pushing the young boy down in his seat. Douglas described anger as a way of regaining control and an attraction to the idea of being a man with power over vulnerable people around him. In the trauma clinics and my forensic work, I had heard so many men (and not a few women) tell me that to be in control of other people, even just for a moment, felt exciting. That desire is nearly always because they feel out of control of themselves, especially if they aren't sure where they end and someone else begins. As children, that feeling is usual; they are semi-permeable, and the boundary between child and parent is blurred. This can mean some parental emotions cross over into children's minds, as with Douglas and his mum sitting side by side on the aeroplane, connected in a shared fear of death.

Later, in adulthood, we become more opaque to each other, and we have to get comfortable with that. I was trying to imagine how it might be for Douglas to realise that he could not control his much-loved son, that it might be time to give him the agency to go out into the world, where Douglas could not protect him. I thought Douglas's unprocessed rage, arising from his experience as a hostage, was getting in the way of helping his son emotionally distinguish himself from his parents and become his own person. Going forward, Douglas would benefit from therapy that helped him relate to emotions of shame and anger differently. If successful, he might be able to loosen his grip on his son, so young Dougie could develop his own sense of masculine agency and control.

Our session time was nearly over. It had been exhausting for both of us, and I wanted to check that he felt okay before he left and travelled home. He was thoughtful, much more settled than when he came in. 'I think it may have done me some good, actually. Even though talking about all this has been hard, some of the things I've said today . . . well, it's a relief to get it out. I feel . . . something's shifted.' He paused. 'I'm sorry if I was rude.' I shook my head to indicate 'no problem', and went on to suggest that it might prove valuable if he tried working with a therapist who specialised in post-traumatic anger management using CBT principles. If he could understand the thoughts that preceded and followed his feelings of rage, he might be able to reflect and contain himself, using deep breathing, muscle relaxation and other techniques that help people feel more in control.

My sessions with Douglas were put on pause while he did that work, and it was almost a year after the 'angry' session before he came to see me again. He looked well, relaxed and tanned, and he was keen to tell me that over the summer, he and the family had flown together for a holiday in Majorca. He and Dougie had hired mountain bikes to explore the beautiful cycle trails, and the family had cooked together and gone sailing. He had also taken up my suggestion that he find a safe time and place to talk with his family about his experience of being a hostage, and he was able to do that with them while on holiday. He needed them – especially his son – to comprehend what he had been through, just as he had once needed his parents to grasp how painful the experience had been for him. He told me it had gone well; he had felt heard and understood.

He still had some anger management sessions to finish. He was also interested in trying some family therapy in future, as his children continued to grow and move away into their own adult lives. It seemed time to end his work with me, as he was clear that he was in a different place and had gained a new perspective on his emotions.

I did want to ask about the airport – how he'd felt in the departure lounge with his family, waiting to board the plane. Douglas smiled. 'You know, I didn't think anything of it. We all sat there together, and I felt safe . . . It seems such a long time since I had that panic attack at the airport, and to be honest, this felt fine. Normal, I mean.' I thought, but did not say, that this is what happens when you've done the work of repair and recovery.

'Think of it this way,' I told him. 'Say you are in pain because of a frozen shoulder, and the physiotherapy work you do to recover is hard. But the absence of pain when you're better is not a loss of something; it's a return to baseline, to where it's good to be.' Douglas no longer needed to panic about becoming 'explosive' on an aeroplane; his feet would take him to check-in, the boarding lounge and on board without difficulty. He'd left the pain in the therapy room, like baggage he didn't need any more. I was happy to wrap up our work together, and he knew where to come if there was a problem in the future.

———

Douglas helped me to understand that both anger and the fear of it can render people speechless, and that therapy gives them the non-judgemental space to express what they need to say. As I write, I think of that young boy sitting with his terror, staring at those bombs strung across the aeroplane's exit door, and of the countless children in war zones around our troubled globe today who are experiencing that same fear of death in silence. Who will help them speak when they can no longer manage their rage?

The Target

The waiting room was sparsely populated, and my eye went immediately to the two young people sitting in a far corner, heads close together, talking intently. I was pretty certain the young man had to be Joel, the name on the referral from the university health service. I didn't know who the woman was but guessed she could be a girlfriend who had come along to give support. I'd been doing some part-time work in this university clinic, after the psychotherapy service at Broadmoor where I had worked for so long was closed in the wake of the NHS cuts. I found I enjoyed working with the students, especially as I was now a parent to two teenagers and developing a new perspective on their growing minds.

Joel and his friend would have caught my eye in any room: she was tall, her braided hair piled high and entwined with some brightly coloured fabric to match her long skirt, while he was of similar height, with close-cropped hair and that indefinable gift of style that somehow makes a simple white T-shirt and jeans look smart. They were a good-looking pair.

In the context of coming to see a psychotherapist, they also stood out due to their brown skin. Professionals in my field have for years struggled to understand why so few people from non-white ethnic groups seek out therapy, compared to the number of white people. It raises the question of institutional racism, as does the comparative absence in the mental health services of non-white psychological therapists.

Back in the 1990s, the murder of Stephen Lawrence (and the subsequent Macpherson Inquiry) forced a national discussion about systemic racism in the UK, which extended from the police force into all health services.[1] Psychiatry came in for particular scrutiny, in part due to the grim over-representation of Black and brown people detained in secure psychiatric services, as well as in our prisons. In 2004, another inquiry, this time into the death by suicide of Daksha Emson (herself a psychiatrist), stated with certainty that psychiatry in the NHS, like the police, was 'institutionally racist'.[2]

Even allowing for differences in data collection and population numbers, there is a decisive imbalance in terms of which ethnic groups seek mental health treatment. In 2014, a survey found that of the 81 per cent of the British population identifying as white, some 15 per cent of them were getting treated for mental health issues, compared to 6 per cent of those among the 3.5 per cent of the population who identified as Black.[3] This could be due to systemic racism; it could also be that the individualist focus of most therapies does not speak to people who come from cultures that are more community-orientated. It is possible, too, that some groups reject mental health explanations for psychological distress and prefer community or religious rationales and approaches to care.[4]

Within the profession today we do see more psychiatrists and psychologists from varied ethnic groups working in psychological therapy services than ever before.[5] Importantly, there is also a heightened awareness that practitioners need to be 'culturally competent' and thoughtful about unconscious bias. I have a long-standing interest in critiques of psychiatry, and I accepted early on in my training that, however well-intentioned, therapists like me might not recognise that they see the world through a lens of white privilege and risk pathologising differences from a 'white norm'. I have found mindful awareness and continued learning from colleagues and patients to be invaluable, and Joel's story would have some essential things to teach me.

As the two young people approached, I introduced myself, my eyes on the man. 'Joel?'

'That's me,' he said with a smile. I turned to the woman with him, who held out her hand, saying, 'I'm Lami.' I invited them both into the office, pulling up a second chair for her. I was still unclear about their relationship and whether Joel was happy she was joining us, but I assumed I would soon learn more. We spent a few minutes getting some background; all I knew from the referral note was that Joel was a twenty-three-year-old history undergraduate at the university. He'd recently been mugged, and there was some question about whether he might be experiencing some post-traumatic stress.

He began by explaining that he had started his course at twenty-one, having taken some time to travel after dropping out of a law degree – 'It wasn't for me.' Lami was a student from Nigeria who was in London on a study-abroad programme, and they lived in the same hall of residence. Joel told me they'd been friends since their first day at uni; she'd become 'like a big sister', he said with a smile. He had invited her along today because she had been with him when he was mugged. Lami nodded in response, adding, 'And I want to be here for Joel.'

I asked if Joel might sketch out his life story a bit before we got on to the details of what happened. He seemed happy to do so. He spoke in an accent much like my own, and I guessed he had grown up in London or the Home Counties, which he soon confirmed, mentioning one of the leafy suburbs of west London. He was the elder of two boys. He said his mother was also born and raised here, but her parents had come to England from Barbados in 1950. She had gone to Oxford to study law, and there she had met

Joel's father, who came from Dublin. 'That's where I get my blue eyes,' he said cheerfully, 'and probably why I thought of studying law.' He said he was close to his parents and younger brother, who was about to start university.

'Thank you, Joel, that's great. Do you think you can tell me a bit about what happened? You were mugged at a railway station, is that right?'

Joel paused, and Lami spoke, her voice deep and accented. 'He was the one they attacked . . . but this is also my story.' I was intrigued that it was Lami and not Joel leading with this. 'It was a racist thing,' she added firmly. I looked to Joel, uncertain whether he needed her to speak on his behalf, and if so, why? He shook his head. 'We don't know that, Lami.' Before she could say more, which I sensed she longed to do, he sat forward. 'Let me tell you what happened, Doctor.' He spoke carefully and deliberately, almost as if presenting a paper in class. It had been six months earlier. 'We were coming back from a play a friend of ours was directing at a pub in Battersea.'

'A bad play. Very boring,' Lami interjected.

Joel smiled, unbothered by her. 'It was pretty late,' he went on, 'past closing time, not many people around. We were going to take the overground back home, and while we were waiting on the platform, some guys sitting on a bench down one end got up and came towards us. I thought they might be wanting to cadge a cigarette or a light, something like that. That was all that went through my mind, honestly.'

As he described the scene, I had a sudden sense of discomfort. I felt my body tensing and hands clenching. What was that about? I couldn't dwell on it – this conversation was about their distress, not mine. I made a mental note to return to it later but had to work hard to bring myself back into the room.

Lami's voice was raised in disbelief. 'Three white men walking towards you on a deserted train platform late at night, and you thought—'

'Lami,' Joel said mildly, 'you'll get your turn.' She slumped back in her chair, arms folded across her chest, frowning.

Lami probably knew the statistics better than I did. Data on violence in peacetime settings, in the UK and elsewhere, show that young men of colour are twice as likely to be victims of violent crime than white males of the same age.[6] One risk factor is perpetrator racism, whereby attacks are motivated solely because of bigoted beliefs about human differences, and especially envy and hatred directed at skin colour.

'She doesn't get it,' Joel said, gesturing towards Lami. His voice and gaze at his friend were warm; he was not dismissing her. 'She's lived her life in a country where white people are a tiny minority, where they're the ones who experience racism.' Lami had been sensitised, he suggested, upon her arrival in England, when she was confronted by overtly racist comments and discrimination for the first time in her life. Before I could turn my head in her direction or begin to ask her what she thought, Lami burst in with 'Tell her about the shopgirl.' And then she proceeded to do so herself. It was her second day in London, and she had decided to go to Chelsea to explore the shops in King's Road. She drifted into a designer boutique and was happily trying on a scarf, when she was abruptly asked to leave by a white assistant, whose manner and language implied Lami was there to shoplift. Furious, she stormed out. It wasn't her last such experience by any means.

Lami finished by saying, 'I don't understand how Joel feels so at home in a place where things like this happen, for no reason, every day. I never felt "Black" until I came to England; I didn't define myself that way. Now, I must watch myself and see myself differently.'

Joel made an impatient gesture. 'That makes no sense.' Lami immediately shot back, 'It does to me!'

'Joel?' I asked, feeling like an ignorant mediator, out of my depth. I was not qualified to guess what made sense in this context. I wondered again if Lami was more distressed than Joel, and why his was the name on the referral.

Joel paused before answering, and his voice was even. 'That doesn't mean the mugging was a racist attack. Sure, I've had my own experiences like Lami's, that kind of everyday racism – being stared at in shops, police slowing down to check me out . . . things like that. Of course.' I was struck by the contrast between his resistance to Lami's thesis and the words 'of course'. 'But I think I've metabolised it, you know? It's just part of being Black in London, but it's a small part, and I choose not to dwell on it.'

I thought his metaphor was important. 'Metabolised? Could you say more?'

Metabolising food breaks it down and turns it into necessary energy; did Joel mean there could be something useful to be taken from negative encounters with racists? A memory floated up from my own days at university: a song whose lyrics describe anger as a kind of energy. Joel sat back, considering my question. I observed him closely, noticing no apparent signs of distress.

'Well, as I said, things happen . . . usually in places where I'm the only Black face. And I can be aware of not being welcome or maybe of not being trusted, but when I open my mouth, sometimes things change again. Sometimes I think I'm being scrutinised for my education and class more than my colour. Class and education, you know how it is, all that "Where did you go to school?" stuff.' He smiled. 'The great British obsession.'

Lami interjected, 'That's the same everywhere, Joel. At home, we are divided into different classes. That's not only a British thing.'

Joel acknowledged that might be true, but continued, 'Sometimes when these things happen, I don't know if I'm seen as Black or white or something else that some people can't understand. I think I've tried, with my parents' help, not to make too much of those experiences. Do I get something out of it? Maybe their intolerance makes me more tolerant of people who aren't like me. I figure racism is their problem, not mine.'

There it was: the 'metabolism' was happening before my eyes, providing an outcome he needed. I briefly thought he could also be telling me this because he did not know me yet or was unsure whether I was of the same mind. He might be trying to regulate me, a Black man reassuring a white stranger by offering an 'I'm okay, so you're okay' message. He sounded so reasonable as he parsed all this out – and then it occurred to me that he might be doing so to avoid talking about being mugged.

'That makes sense. Thanks, Joel. Now, are you okay to go back to what happened on that train platform?'

'Oh. Right. It all happened so fast. I mean, these guys came up to me and said something I couldn't hear, and I guess I said something back, like, "I beg your pardon?" And that seemed to make them angry. One of them copied me, saying "Pardon?" in this over-the-top posh voice. And then he stepped forward and pushed me in the chest with the palm of his hand. And then Lami shouted something like "Hey" or "Stop" . . . right?' He glanced at her for affirmation, and she nodded. 'I think I might have put my hands up,' Joel said, 'not to push him back but to calm him, you know, sort of "Take it easy, man." That's when he slapped my hands down and pushed me to my knees, and then his friend tried to kick me, but I moved, and he missed.'

'And then?'

'My jacket fell open, so there was my wallet in easy reach, and one of them grabbed it just as a train pulled in – and then they ran off, thank God.'

'And then?' To my ear, my voice sounded a little sharp, and again I experienced a twinge of something, an unease that I could neither name nor completely ignore.

He looked at me, a little puzzled. 'That was it.' I had heard language like this before, where the speaker implies that the trauma is over as soon as the immediate danger has gone. I knew that wasn't always true, but I wondered if Joel wanted it to be the case.

'Tell her what they called you,' said Lami abruptly, breaking her self-imposed silence.

'Do *you* want to tell me?' I asked her.

Joel sighed. 'Go on.'

She sat up in her chair, leaning forward. 'He forgets the most important thing. When that bastard tried to kick him, he called him a—'

'Cunt,' said Joel wearily, turning the tables and becoming the interrupter. 'So what?'

'That's not it,' Lami shot back.

'I was there. He said it right to my face.'

I was confused. Lami was furious, and I wasn't sure why.

'Lami?'

She looked at me, brows low. 'We do not agree. I heard what I heard. The man said, "Coon."'

I knew this racist slang, of course, though I would probably have associated it with American diction. Everything eventually makes its way over the sea to us, for good and ill. Lami's reference also reminded me that people's experience of racist trauma and victimisation may be subtly – or not so subtly – different in Britain and the US, and indeed elsewhere.[7]

Still, one thing language slurs do have in common is that they are intended to demean and dehumanise, and I was interested that these two young people had heard things so differently. Was it possible that Joel might *prefer* to be called a 'cunt', instead of an explicitly racist term? It occurred to me that anyone can be on the receiving end of that word, a powerful term that doesn't refer to how people look but can be used to degrade both men and women in different ways. In contrast, 'coon' is specifically a racist word denoting a person's skin colour. I wondered if Lami could see this more clearly than Joel because of her experiences since coming to England from Nigeria. In using such language (if he had), the white mugger had attacked Joel's Blackness, just as Lami's had been denigrated when she'd been accused of shoplifting.

A colleague of mine talks about the paradoxes of racism: it is random yet ubiquitous; profoundly personal and also impersonal, because it ignores an individual's life experiences and identity. He likens the impact of a racist slur, even though it often comes from someone you don't know at all, to the intimate pain of a person you love using things you've told them in private against you in a moment of anger, saying hurtful things that can never be taken back. This was important for me to consider in the context of post-traumatic stress because of the evidence that traumas caused by humans tend to have more impact than other kinds, such as transport disasters. The survivors of the helicopter crash I'd met didn't feel that the machine had had malicious intent towards them. In contrast, those who are the victims of assault (verbal or physical), such as Douglas, the man taken hostage by hijackers, experience a bewildering sense of being both the subject of intense malice and a person whom the attacker doesn't care about at all. I was also reminded of a man who raped a woman, saying to her as she protested, 'I'm the boss now, and you have to do what I say.' There was no recourse; she was not only physically violated, but also robbed of all agency.

I looked at Joel. 'What do you think about what Lami's said?' I didn't want to correct his version of the experience, but I couldn't gloss over this question and move on if more space was needed to explore what she'd said. After all, he'd brought her with him, suggesting he needed her there. But he shook his head, and his voice when he spoke was almost irritable. 'Whatever. That's not what I thought.' He turned to me, making full eye contact. 'I'm not sure if we're wasting your time here, you know?'

It's always interesting when people say this during an assessment, and it's usually a signal that the speaker wants to escape some painful feeling. I didn't respond directly; instead, I said I felt I was getting a sense of how he and Lami had experienced the assault differently. 'I can hear that this was a really nasty and frightening experience. Maybe you are both "metabolising it" from your different perspectives? Have either of you had any symptoms of stress or distress that have been bothering you since it happened?' Given that they'd come to a trauma counselling service together, I thought I needed to check this out. Again, I was conscious that Joel had been named in the referral, not her, but it was possible that Lami could be the more distressed of the two yet had somehow been overlooked – or maybe she was deliberately avoiding scrutiny.

Joel shrugged his shoulders. 'I'm not sure. Coming here was Lami's idea, really.'

This was important. But before I could follow up, Lami broke in again. 'It's a trauma! They could have killed you!'

Joel shook his head. 'I don't know . . . I wasn't hurt. I was frightened at that moment, sure, but now I feel fine. What good does it do, going over and over the story again? People perceive the same events differently; we talk about this all day in our history seminars, don't we? Even if they were racists, at the end of the day, I'm all right. They took my wallet; that was a pain, cancelling cards and so on. And yes, it was scary, but nobody died. What are we doing here?' He broke off, gesturing at the door as if to take in the whole clinic. 'Why are we calling this "trauma"? Why don't we call it "something upsetting"? Or "a bad night out"?'

'Here he goes,' Lami said.

I couldn't argue with what he was saying, and I was still aware of feeling uncomfortable. I was distracted, too, as Joel's question had gone right to the heart of one of the biggest debates about trauma treatment: what *counts* as trauma? And relatedly, who gets to decide how 'trauma' differs from all the other upsets in our lives? I knew that the presence or absence of physical harm was not decisive; I'd worked with other patients who'd had similar encounters with street violence or muggings, where 'nobody died', as Joel put it, but who later suffered from acute PTSD symptoms nonetheless. Some research was emerging that confirmed something I'd observed for years: the subjective meaning of a traumatic event to the victim can be associated with the level of distress they feel, regardless of any physical threat or harm.[8]

Joel and Lami's opposing viewpoints highlighted another essential question: are there some kinds of violence that *must* always count as traumatic, such as sexual assaults or racist violence? What if that label doesn't feel right to the victim? And did it matter what Joel's assailants intended or how they felt about his skin colour? Bluntly, was it 'better' (whatever that means) to be called one c-word over the other? And who gets to decide? These were not questions I could answer as a psychiatrist; it was unclear whether they were even medical questions. What I did know is that people use language as a weapon to shame and intimidate others. Who can forget the great childhood lie that 'Sticks and stones may break my bones, but words will never hurt me'? Words may get under our defences as surely as a sharp blade, and physical assaults often start with slurs and swearing as a way to intimidate and demean.

I was trying to work out where to go next. So far, Joel had not mentioned anything about feeling mentally or physically affected by the mugging. Although he could be avoiding his distress, that did not seem to be the case. He was not saying he was unaffected, only that he was not ill. If that was his view, I was keen to avoid medicalising a situation he was managing well, according to his values and perspectives. Meanwhile, here was Lami, who seemed to need her friend to say that he had been harmed and to call out racism as an injury; perhaps unusually in their friendship, they disagreed on something profound. It seemed as if her sense of Blackness differed from his, which was understandable: he was mixed heritage and had grown up in a multi-ethnic but majority white country, and she had lived all her life up to this point in Nigeria. But I was also mindful that my understanding was superficial; I wasn't remotely qualified to assume anything about what their Black identity meant to them. And I was not clear with regard to who was suffering the most here, though I felt uneasy about assigning Lami a role she hadn't sought.

We were coming to the end of this assessment. I decided to leave the next step entirely up to them. 'What would you like to do? You've both raised good questions about how to understand an experience like this. I don't think it's ever a waste of time to talk things through – to digest them properly, if you like. But I also understand that you see things differently, even though you were together. I suggest that you go away and think about it, and if either or both of you would like to meet me again to talk some more, call the clinic. I'll tell the admin team. And I'd like to suggest that you come separately, if you want to talk. How does that sound?'

Lami looked doubtful, maybe even a bit defensive. 'Don't you think Joel needs treatment of some kind? He went through something bad.'

'He did,' I said. 'And so did you. You witnessed a violent assault on someone you care for, and you didn't know whether you might be next. It must have been really hard. That's why it might be a good idea for you each to have your own space in which to talk with me.'

Lami relaxed a little, and as they left, she thanked me, gripping my hand firmly, before leading the way out of the room. At the door, Joel turned back briefly to clarify: 'So, we don't *have* to come back?'

I looked him in the eye, choosing my words carefully. 'I don't think that either of you *have* to. But as I said, it could be useful to talk about it, if you *want* to. Take some time to think about it. And it can be whenever you like. No pressure.' I made sure my tone was warm and empathic. 'I think, given what was imposed on you both, it's important that you are the ones to choose what you want to do next, to decide what is right for you.'

After the session, I realised I was still feeling uneasy, tense, and I needed some air. It was lunchtime, so I took a walk around the block. Only when I passed a Tube station entrance did the penny drop. I stopped on the pavement, allowing passers-by to jostle me, my eyes fixed on the staircase descending into the station. Joel and Lami's experience had direct parallels with something that had happened to me three decades earlier. I hadn't thought of it for years, but here it was now, as live in my mind as if it had occurred the day before.

———

I was running with my boyfriend Ash to catch the last Tube. We'd been to a concert at the Royal Festival Hall, having heard some glorious music courtesy of our student-price tickets. I remembered our laughter as we raced down the stairs, jumping onto the train just as the doors were closing. We had our pick of the seats; we were alone but for a few young white men huddled together at the far end of the carriage. Ash was a beautiful man of mixed Spanish and Indian heritage, a musician and fellow medical student. As the train rushed to its next stop, we sat close together, not speaking as we recovered our breath.

I'm not sure when I realised that the men down at the other end were laughing and glancing our way, nudging each other. One of them, sporting some wispy facial hair and a wool beanie hat, drained a beer can, dropped it and gave it a kick, so that it rolled along the carriage and past our feet. Then he and one of his friends moved closer, looming over us, and he said something unintelligible, his voice low but hostile and menacing. He confronted Ash, a finger in his face, as if trying to provoke him, demanding, 'What you looking at?' Then, leaning in, he added, 'What you want, you fucking cunt?' Ash made a ducking motion with his head, as if to shake him off, and I sat numbly, holding my breath.

As the train doors wheezed slowly open at the next station, the one who had been speaking to Ash abruptly raised his hand and hit him sharply across the face, one–two, one–two, before jumping off the train with his friends, shouting and laughing as they disappeared down the platform. I remember feeling stunned as I looked at Ash and saw blood trickling from a cut across the bridge of his nose; his assailant must have been wearing a ring. I turned his face towards me so I could see the wound. It seemed no more than a deep scratch, but there was a fair bit of blood. 'Are you okay?' I asked. He nodded, wordlessly. I took his hand. 'Do you want to go to Cas [medical-student shorthand for Casualty]?' Ash shook his head. 'Let's just go home. I'm okay, Gwen.'

We never told anyone about the events of that night. I think we briefly considered reporting it to the police, but what could they do? We had hardly seen the men, so identifying them would have been impossible. We had no idea why it had happened either; it had come without warning or context. I think we both thought about whether Ash had been the target of racism, but that didn't help us make sense of what had happened. Even then, with no forensic experience, I understood that we were just in the wrong place at the wrong time with the wrong people in the wrong state of mind.

––––––

As I stood there outside a Tube station many years later, I recognised what was happening. As so many of my patients had described to me, I was having a memory flashback, not unlike the literary or film device whereby a memory or event from the past interrupts the forward progress of a narrative. Another person's story had reactivated memories of that Tube journey with Ash, bringing old feelings of discomfort and anxiety to the forefront of my mind. I think I felt some surprise: flashbacks are thought to be unique to PTSD,[9] and I'd never considered myself as having a residual post-traumatic problem; nor had Ash, as far as I knew.

From time to time over the years, if I was asked about past fear experiences, I might have recalled that episode with Ash, but generally, it had never intruded into my thoughts unbidden, until that day when Joel and Lami brought their story to me at the clinic. Now I understood why I had felt so uneasy when Joel told his story. I considered whether I might be feeling some of his anxiety for him; emotions can be contagious. This was neither the first nor the last time that I would see how an odd response to a patient's story would turn out to hold meaning for their therapist – something particular to them.

————

A week or two passed. I had a hunch that Lami would contact the clinic, because her anger and distress had been so evident. I thought it was possible that Joel, on the other hand, might want to dismiss his experience, much as I had done with mine once, ascribing it to 'life in London' and moving on. However, there was no follow-up from either of them, and I began to think they had not seen the service as helpful. Maybe Lami had projected her feelings of victimisation onto Joel and did not wish to expose herself further, especially not to a white woman who might misread her, as the shop assistant in Chelsea had done.

As for Joel, I was reminded of the work I'd done in the trauma clinics with refugees from war and asylum seekers. They, too, had resisted the medicalisation of their distress and the idea of being labelled as 'traumatised', as if something was wrong with *them* rather than with the perpetrators of the brutal repression and violence from whom they had fled. They taught me that the meaning of an experience of discrimination or societal violence, and the decision about whether treatment is needed, must rest with the survivor, not with their doctor.

With the benefit of hindsight, I could see that following our experience on the Tube, Ash and I had also resisted becoming identified as patients or people in need. Part of the reason for that may have been that our identities as doctors were in formation, and many medical professionals do not want to be seen as patients. But the reality was that I *hadn't* had any serious post-event anxiety, certainly not enough to get in the way of everyday life, which is the medical and legal measure of post-traumatic distress. It also occurred to me that I might have rejected victim status because that would have given some power to the perpetrators. Maybe that could be the case for Joel, too.

So I was intrigued when, some weeks later, Joel did call the clinic to make another appointment to see me. I was curious to hear what he would say. He arrived on time and seemed much as he was before, measured and calm, speaking spontaneously, without hesitation. He began by showing me a well-known text about PTSD that he'd borrowed from the library, fishing it out of his backpack and handing it over, many of its pages marked with slips of paper, as if he were researching an essay on the topic. He'd made a note of the most common symptoms and ruled them all out, ticking them off on his fingers. He wasn't irritable, wasn't experiencing flashbacks and didn't have nightmares. But—

'But what?' I prompted.

'Well, I couldn't find this anywhere, but I wanted to ask you: I'm getting this kind of . . . I don't know what you'd call them. I'll be in class, and sometimes I'm tuning out what's happening around me and thinking about those men again, about the mugging. But it's not a flashback, not as they describe it here, anyway.'

I emphasised that although the word is widely used, flashbacks are still not well understood, nor are they experienced in the same way by everyone who has them.

'But they always have a trigger, don't they?'

I admitted I'd come to dislike the word 'trigger'. In the trauma context, it does not do justice to the different ways people remember, or to what meaning past events hold for them as individuals, at various stages of life. There is a popular misconception that memory is an armoury of weapons, locked and loaded, ready to fire at any provocation; or a computer file containing a record of everything that has ever happened to us, which can pop up on our mental screen at any time, unbidden. Both metaphors belie the awesome complexity of the human mind. I see memory of past distress as more like an alarm, drawing attention to events that we're not otherwise conscious of. This is what happened to me after my previous meeting with Joel. I didn't mention my own experience to him, but I did say that memories sometimes flag potential threats. I added that they can also illuminate things, helping us gain clarity when our view is obscured and get a better understanding of why we're feeling 'inexplicable' discomfort and distress.

Joel nodded. 'I get that. I don't feel "triggered", though,' he said, putting the word in air quotes. 'I'm going about, living my life normally. I don't avoid going on public transport . . .'

'But something is happening, you said?'

'Yeah. I want to call them daydreams, but that sounds . . . a bit fluffy somehow?' He was choosing his words with care. 'I think "daymares" might be a better word, although I'm fully awake. I find myself running over what happened in my mind, but I'm changing the script, or trying to, like you do in a nightmare sometimes. You know it's a bad dream while you're having it, so you try to force yourself to wake up just when the monster is about to get you.'

'Monster' sounded like an important word in this context, and I was interested to know what it signified to Joel, but that could wait. Instead, I asked, 'Can you describe how that works, "changing the script"?'

'Okay. So yesterday, I was in class, and I started thinking about how that guy was mimicking me, saying "Pardon?", all sarcastic, and how he pushed me in the chest. But then what I saw . . . no, not *saw*, but let's say . . . imagined . . . was that he pushed Lami, too. Which never happened.'

'Go on.'

'Lami starts crying, and I get so mad I go for him – and it's so real, Doctor. I mean, I can feel the leather of his jacket under my hand. When I shove him, the guy falls backwards, and his mates race off. Then I'm kicking him, hard, in the ribs, over and over. He's all doubled up in a ball, and I keep kicking till he's begging for mercy, apologising . . . And there's Lami, lying on the ground, but she's encouraging me, too, like, "You show him, Joel!" Which she'd never do in real life, not in a million years. She'd be right in there, kicking him, too, or dragging me off – one or the other.'

'And then?'

'Then the bell rang for the end of the lecture, bringing me back into the room.'

I sat in silence, contemplating his experience of this alternative reality.

'It's happening a lot, every few days or so. Have you ever seen this before?' Before I could say anything, he went on, insisting that this wasn't like him, that 'it made no sense'. I could tell there was real suffering for him in that idea of the irrational, and I understood why he had come back to see me. He was a rational young man, and making sense of things was central to his course of study and seemingly to his peace of mind. Long before the invention of psychiatry, people would seek out those who could 'provide sense' of distressing events, be they shamans, priests or poets and playwrights. As Viktor Frankl knew well, the search for meaning is a driving force in all human beings, especially when they are confronted with irrational violence. How anxious was Joel about losing his sense of reality altogether? That is one definition of madness.

'I'm a pacifist,' Joel was saying. 'I mean, I've never been in a fight of any kind, not even as a kid, never – and it's all nonsense anyway. I couldn't have taken this guy on; he was twice my size, big and mean. It was just . . .' He stopped.

'A daymare,' I said thoughtfully.

'An invention. But why would my brain do that? Have you had other people who have experienced anything like this?'

I had to respond to that repeated anxiety. 'I have seen something like it, yes . . . It sounds like a painful preoccupation with events you still struggle to understand. Perhaps your mind is generating images of something that would make more sense to you?' Or something that might be more appealing, I thought to myself but didn't say.

Joel nodded eagerly. 'I get that.'

'Shall we look at the different aspects? Maybe start with your images of Lami in distress, needing your protection. Does that sound real to you?'

Joel chuckled. 'Absolutely not!'

'But in your daymare' – that seemed an okay word to use – 'you are the hero, confronting the bad guys, like in a cowboy film.' Even as I said it, I wondered if he would see this comment as rooted in the language of colour, considering those films always used black and white hats to denote who was heroic and who was not. But he just added his own, more generationally apposite take: 'Or like Batman.' He grinned.

I continued. 'So, in this version, against all odds, the guys who threatened you are beaten and run away. You have all the power and agency; you are the man of courage, the good guy who makes bad men feel ashamed. But that makes me wonder if you have been left with some shame, too. It's a powerful emotion, a judgement we make of ourselves when we think others might accuse us of failing to do something we should have done.'

Joel nodded, thoughtful. And then I added, 'It contrasts with something you suggested last time we met.'

'How do you mean?'

'When we first met, you mentioned some experiences of casual racial abuse that you had chosen to either ignore or learn from?'

Joel looked uncomfortable. 'Sure, but I'm still not convinced that the mugging was about race. I mean, for Lami, it's always about race, and I get that . . . but I genuinely don't know what name the guy called me, and if I'm honest, I feel irritated that Lami immediately makes me out to be a Black victim of a white oppressor.'

'But in your daymare,' I pointed out, 'you are not a victim. I'm not disagreeing with you, Joel. I'm curious about how your wish not to be another Black victim fits with where your mind is taking you?'

Joel sighed. 'I just feel it's not that simple . . . not for me. I don't default to seeing myself as Black among whites, you know? Sometimes I feel more Black or brown than Irish, but other times it can be the other way around. Depends on the context.'

I thought I understood what he meant now. I'd heard others speak about this idea of a chameleonic racial identity, expressing a disconnect with the polarised public discourse about race. I also sensed something in Joel that was glad or even proud of the fact that he wasn't as quick as Lami to assume he knew what the assailant had called him.

'Why can't I see myself as both white and Black? I don't feel like I have to choose. I don't *want* to choose.'

He sounded a little angry, and I asked if that was the case. 'I just hate that those men are intruding into my world, and I hate that they got to decide what to call me, whatever c-word they liked . . . That's the thing I wish I could put right. I don't have any decent recourse . . . there's no linguistic equivalent for me to throw back at them, is there? And they get to write me off in just one word.' He'd summed it up: racial slurs silence the recipient, annihilating all defence.

I didn't speak, hoping the silence would prompt him to say more, and it did. 'So yeah, I'm angry . . . and I think I also feel ashamed that I couldn't stop them, that they defined me as a victim, even for a minute. I hate to admit it, but maybe it's about gender, too? Like, I prefer my made-up story, where they pick on Lami as the victim, and I step in to protect her. I'm the one that gets to do the shaming.'

'Does that help explain why you are kicking someone on the ground in your daymare, like they tried to kick you? You could have been badly hurt, but perhaps the new script gives you a chance to be the one who hurts people who try to hurt others. A kind of avenger?'

Joel grimaced. 'Ugh. I don't want to be that guy. I mean, it just makes me like them – a monster.' There was that word again. I thought of its root, *monstrare*, 'to show'. What was being revealed to Joel in his daymare? A part of himself he'd rather not see? I studied him, taking in his evident distress and intense desire to work this out. So often, the answer to our problems lies in the stories we tell ourselves, I suggested, the ones that help us make sense of the senseless. 'You're making me think about those myths and fables about a young hero who goes on a quest and must slay several monsters before he returns home. In our real lives, the biggest monster is often the one inside us. I wonder whether your new script is a way of managing big emotions that don't fit with how you see yourself? They sound pretty human to me, but maybe they feel monstrous to you?'

'I don't like feeling ashamed, and I don't like feeling angry. That's all. I'm not an angry person.' He avoided my eyes, looking out of the window to my left. I felt his unease. 'If I'm attacking this white guy, doesn't that make me as bad as him?' Then he turned to face me and blurted out a confession: 'Doctor, I have to tell you, the last time I had one of these things, these daymares . . . it ended with a train coming in, and me pushing the guy onto the tracks, right into its path.'

I saw the horror in his eyes as he imagined what came next. After a pause, I offered gently, 'Perhaps, just for a moment, you had a sense of what it might be like to be that person who mugged you, who had a mind full of hatred and a wish to hurt? Who was angry about something that we can't know?'

'Right . . .' Joel said. 'But *I* am the one that should get to be angry, not them! It was totally unprovoked! What the fuck did I ever do to them?' It was the first time I'd heard him raise his voice or express real anger. I was willing him to go further, and he did. I braced myself as he shifted into a higher gear. 'I'm fucking furious at those little shits, okay?' He smashed his fist down on the table, then winced. 'Ow.' He laughed at himself, rubbing the side of his hand. 'I'm not much good at this. If Lami was here—' He stopped short.

'Go on,' I prompted. 'What if Lami was here?'

Joel gave a heavy sigh. 'She'd say I was afraid of being the angry Black man. And I get it, I know what she means. You hear that cliché a lot.' I thought of the writer James Baldwin and quoted his comment that 'to be a Negro [in America] and to be relatively conscious is to be in a rage almost all the time'.[10] Joel waved that idea away. 'Yeah, I've heard that. But I don't think it's right. Not for me. At least, I want to be angry, without it having to be put in a box marked "race" or "colour". It's about me and my experience. Nobody gets to name that except me.'

We sat quietly for a few minutes, digesting this. Metabolising it, maybe.

'We all need someone to hate and to fight against,' a Black colleague of mine wrote to me recently, 'in order to rid ourselves of something; maybe racism is less about power and more about this.' I wish I'd had that idea back then to share with Joel.

He broke the silence. 'What do you think I should do now?'

I took a moment to reply. 'I don't know that you "should" do anything.' I recalled that I had said this to Joel the first time we met. 'It sounds like you are working on something important about your experience. Making sense of events like this takes the time it takes, there's no schedule for it, and I guess your daymares are your way of doing that. It's not abnormal or unusual, and the content of those daymares seems to raise some issues beyond this single event – things to do with identity, colour and gender and your sense of self, on your terms. I'm wondering if our trauma counselling service is the right place for you to explore what you're feeling. You can come back and talk with me some more, if you like, but I think you might get more out of talking to a therapist who can take a broader and deeper perspective.'

'What do you mean?'

'Well, most individual or group therapists work with people who have had distressing experiences of different kinds; it's what we do. We also see people who are exploring their identity, and as I'm listening to you, I'm hearing that this experience has raised some questions about that for you, which might be as important as the mugging itself.'

Joel looked taken aback. 'Oh. I hadn't thought about it like that.'

'And given our discussion is partly about race, is seeing a white psychotherapist an issue?'

'Wow.' Joel considered this for a moment, as did I. I think both of us were weighing up which differences 'count' in the therapeutic relationship, which is not an easy thing to answer. When I'm working, I think things like my age, gender, heritage or skin colour matter far less than my identity as a therapist, because that's what will enable me to hear someone out and reflect on what I struggle to absorb. Perceptions of sameness and difference are part of how we relate as humans; doctors have never had to be an identical match to their patients in order to help them. For me, the question is always whether it's possible to form an alliance with someone that is a basis for trust and curiosity about themselves. The work is about them, not me.

On the other hand, I know that feeling understood and comfortable enables that alliance to be a good one, and that immediately raised another question in my mind, related to how all this began – with him and Lami, back on that train platform. 'Maybe the issue, Joel, could be that you feel vulnerable being watched by a woman as you work through a problem. Is that something to think about?' I smiled, wanting to ensure he didn't see this comment as a rejection of him, a closing door.

He met my eye. 'I don't know. But I don't want to start over with someone new . . . and I still don't know if I have an actual problem. Can I think about it some more and maybe get in touch again?' I told him that sounded sensible, and we left it at that.

A couple of months passed, and when Joel asked to come back to see me, he did not begin by talking about the mugging or his daymares, but instead about his studies. He was struggling and not doing well with his coursework. He was meant to be choosing a topic for his final dissertation, but he was finding it hard. 'I keep changing my mind,' he said, frustrated. 'My tutor is giving me extra time because of the whole . . . you know.'

'The mugging?'

'Yeah. Which only makes me feel worse, because I don't want special treatment. There's nothing wrong with me!'

We sat there, neither of us giving voice to the obvious: he wouldn't be sitting with me if that were so. I thought of something that had come up briefly in our first session. 'Can I ask, Joel, what made you switch your studies from law to history?'

'My parents are both lawyers. I think I told you. Mum does employment law, Dad's more corporate. They always wanted me to follow in their footsteps, and I wanted to be like them. But it just wasn't for me.' He shrugged. 'It seemed so dry, lifeless even.' I nodded, waiting for him to continue. 'Mum and Dad were probably disappointed, but they accepted it – and then I didn't know what I wanted to do. I temped in an office for six months, and then went to Barbados to see my mum's family. You ever been?' I said I had – it was a place I loved to visit.

'I love the history there,' he continued, 'the mash-up of different stories and ways to be Bajan . . . There's the British colonialism, the sugar plantations and the slave trade and all that, but there's also the history of Barbados in its own right, before and since independence – its democracy, all the politics around the tourist trade and the way the community is made up of different groups. I worked in a hotel dealing with white tourists who assumed I was a Black Bajan, while the local Bajans said I sounded like a tourist because of my accent, and I couldn't adapt my voice to fit in. But as it turned out, it was a lovely time. I got to hang out with a mix of people and talked a lot with my grandparents, cousins and aunties. It was so different from life in west London, and I felt right at home, even if I did sound different. I felt like my mind was expanding, as was my sense of who I was.'

He saved up enough money to go travelling in the US. 'That was weird. Totally different again. I started in Florida, then went on to Texas and some other Southern states . . .' He paused. 'Well, maybe you can guess what I'm going to say. I ended up in Alabama. I wanted to look at the history of the civil rights movement. I started to think that maybe I would study that when I returned to uni. But I ended up staying there only a few days. Don't get me wrong, no one was horrible to me, but it felt strange; the atmosphere was sort of . . . thick and tense in some places. Worse than anything I'd known before. It was unspoken, implicit? Honestly, I couldn't always be sure if it was there. I just know I felt uncomfortable about it the whole time.'

'It?'

'The prejudice. It felt like a constant – and maybe it was all mixed up with being a visitor and a tourist and sounding weird to them, like in Barbados; my English accent immediately marked me out as alien. And that's what I felt: alien wherever I went. Did you know that's what they call people in the States who aren't citizens, even if they've lived there for years? "Resident aliens".' He laughed briefly. 'Mad, isn't it?' His comment was so interesting to me, given his prior mention of monsters; a monster is a type of alien, too.

'My dad has some family in New York, so I decided to go up there, and that was more like London – multicultural, normal. Yeah, I felt normal again there, as I do here.'

He changed tack abruptly, turning to the mugging without my prompting. 'But then those guys at the station . . .'

I'd been waiting for this – for Joel to make his own connections.

'The guy that hit me . . . he robbed me of that normal feeling, and it hurt me in here.' He pointed at his chest. 'Not physically but, you know, my pride. He took my right to name myself, and I couldn't fight back.'

I thought immediately of Shakespeare's line in *Othello*, when Iago speaks of how one's name and reputation cannot be stolen like money.[11] I also remembered how slave owners would wipe out people's identities by assigning them a new name without consent. Shame at the loss of identity can easily drive people to violence, especially men, but this was what Joel sought to resist.

'You couldn't fight back,' I echoed, returning to Joel's train of thought.

'Yeah. They had the power. Like you said when we talked about my daymares, I felt ashamed, like I had in front of Lami.'

'Do you think that's what Lami was picking up on when she got you to come here? A sense that you felt vulnerable in front of her and didn't like it?'

Joel nodded. 'Maybe. She's a good friend, and she's sensitive. Maybe she saw me as injured. But more than anything, I think she wanted to register the outrage of it and make sure it was not just another episode of racist violence that didn't get reported or noticed by anyone. And I understand that. But by needing to see a doctor, I still think that means they've won in some way. And I don't want that.' He made a sound of frustration and impatience. 'It's been months. Surely by now I should be over this. Why can't I move on?'

I tried to think of the best way to answer this. 'You talked about metabolising stress, Joel, when we first met – including stress that's racist in nature. When we metabolise food that dis-agrees with us, we can feel sick – but then we get better as the unpleasant stuff passes through our system. Maybe this experi-ence just needed more time and space to move through you because of the emotions of rage and shame it aroused?' And I had another idea to offer: 'Maybe you needed to have a witness, too, someone who isn't Lami, who can keep you company while you work through this?'

'Someone who isn't Black?'

I didn't answer this, hoping he would do so himself, which he did soon enough.

'Someone objective? Or different to me?'

'And female.' I couldn't tell how significant that might be, but I thought it was worth marking.

'Well, yeah. Oh – did I tell you about Lami, by the way?'

She had decided to leave England, return to Nigeria and finish her studies there. I think I felt some surprise about this, and asked if he thought their shared experience had influenced her decision.

Joel frowned, unsure. 'I think she misses home, really. But maybe.'

Maybe, I ventured, she wanted to return to a place where she felt she wouldn't be attacked for something she couldn't change, where her skin colour did not mark her as a target.

Joel thought about that for a minute. 'I think Lami may have been much more upset about it than she could show you or even me . . . Even though she's quite a talker, it seemed like she didn't know how to talk about it. And, yeah, maybe she wanted to be back where her skin colour was something she didn't have to think about. Whereas I'm used to the ambiguities of it – it's been my life.'

'Joel, I don't have your experience, but listening to you, it does sound like being comfortable with ambiguities is a real strength for you. You get to choose your resolution. And what you say helps me to understand why you might resist being put in a box based on your skin colour or profession. Identity, personhood and ambiguity – something for us to think more about together?'

He grinned. 'Definitely.' For a moment, I saw Joel as the eager student of his own mind who was glad to have landed in a class where he might debate and discover more.

———

He returned the following week, and we continued to have more thoughtful conversations. As the weeks passed, Joel reported that his daymares were fading as he began to think about his anger and shame in a different way. He recognised that he had been trying to assert control over painful feelings of helplessness and impotent rage, while at the same time striving to be a 'balanced' guy.

It still felt like he was anxious about being angry, but it wasn't clear that talking with me would help. What was clear to me was that once again, as with so many people I'd seen after a traumatic event, the language of diagnosis and PTSD was not always helpful. I could not resolve the impact of racist abuse for him, much as he had not been able to decide which c-word had been hurled at him. He was stuck between two types of rage somehow, and I had a growing sense that the only way to resolve things would be for him to have some therapy, either individual or group. I raised this with him again, commenting that in therapy, he could explore issues of identity and masculinity in a broader context, rather than keeping a narrow focus on his story of trauma and victimisation. Joel agreed that he would like to explore this, and I gave him some names of therapists to contact.

We made one final appointment. I could tell something was afoot as soon as he arrived that day. His mood was bright, his energy high. He had some news for me and was bursting to tell it.

'I've got an idea about my dissertation . . . and it's really down to coming here and talking with you.'

'How so?'

'I had been planning to write about the Black experience in Britain in the 1960s, maybe interviewing people like my mum's family about their experiences when they migrated here, how they experienced racism in a new cultural context. But now, maybe I want to go further and look at how the language of mental health ties in with this. I could ask them whether they ever saw themselves as needing professional help or as being unwell. Then or now. I need to read up on this and talk to my tutor . . .'

I could see his mind moving into intellectual, analysing mode. I had an image of his distress being poured into a crucible and transformed by the heat of words and ideas, one mental process bringing order to another. I told him I thought he was fortunate to be able to use his mind to help him manage his response to trauma in this way, and I thought it sounded like a fruitful course of research. Years ago, after my experience with Ash, I suppose I, too, had used my training, intellect and capacity for questioning to get to grips with the fear and rage I felt about that random incident on the train. Not long after that, I had gone on to choose my specialism, deciding to go into forensic psychiatry. Maybe this was my version of transforming fear into understanding and making good. Something I thought of rarely and had left behind long ago may have changed the course of my life.

Joel's animation brought me back to the room. 'Something's shifting in my generation,' he was saying. 'Just look at Lami and compare her response with women of my mother's generation.

My mother would have gone to the police or some civil rights organisation and tried to get the mugger arrested, but Lami's first thought was to get me to this clinic. And my Irish grandmother would probably never have told anyone if she'd been racially abused.'

I was glad for Joel; I felt grateful for all that he had taught me. Now, in the autumn of my career, I realise that the best people to learn from are those 'experts by experience' who come and take a risk by sharing something of their thoughts. Joel's searching questions validated my own. He was transforming his experience into a deeper enquiry, and I felt privileged to be a part of it. My identity has been shaped by many experiences, not just the distressing ones, and I knew the same would be true for Joel. He had so much to offer.

It was nearly time for Joel to go. He stood up and shook my hand. 'It's been good to talk to you about all this. I was upset and angry, even if I didn't realise it for a while . . . and it has helped me to say these things out loud, to have a witness . . .' He grinned. 'I love that.' He changed his voice, his accent becoming Southern American. 'Ladies and gentlemen, can I get a witness?' I looked at him, puzzled for a moment, and he explained, in his normal voice, 'You know, call and response, like in those Pentecostal churches when people jump up and testify.'

'Of course,' I said, thinking about James Baldwin again, who was raised in that environment.

'I know Lami was my actual witness on the day,' Joel said, 'but I think I needed someone who was not invested or outraged. So thank you, Dr Adshead. Is it okay to say I hope never to see you again?'

I laughed. 'Absolutely! Psychiatrists are just like other doctors: when our work is done, we hope it's done for good.'

After he'd gone, I reflected on what I'd learned from Joel and what I would take away 'for good'. These kinds of encounters are gifts, and I was glad that he departed feeling positive. But a question lingered in my mind about whether I had failed Lami in some way by not connecting with her and helping her, as I had tried to do with Joel. She had carried her distress back home to Africa, and I would never know what that might mean for her still-developing mind. I found the idea disturbing.

The Trainee

I was ready at my desk when the familiar sing-song Zoom tone sounded, and I immediately clicked on 'Join', relaxing my shoulders and settling in my chair as I waited for a face to resolve on the screen, next to my own. After a second, Ray appeared. He's a warm, open-faced man of fifty or so who already sports a head of silvery white hair. Seeing his lips move, I assumed he was giving his usual cheery greeting, but I couldn't hear anything.

'Hi, Ray. You're on mute . . . ?'

'Oops, sorry, Gwen. Hear me now?'

His deep voice filled my headphones, and I reached to adjust the sound slightly. 'Loud and clear,' I smiled.

In my semi-retirement, I offer supervision to other consultant psychotherapists. They seek it partly as a professional requirement, but good therapists also know they need spaces where they can talk about their patients and do some self-reflection. I've been fortunate to have some excellent supervisors myself, and I'm always glad to be of service in this way. I see everyone online, usually monthly. Ray is one of my regulars, a former GP now working as a group therapist in London.

Although he's not a forensic psychiatrist, Ray and I have much in common. He has worked in the prison service, has a long-standing interest in forensic practice and he shares my love of group therapy. Also, we're both deeply concerned about the effects of the drastic cuts to mental health resources in the NHS in recent years and the grave toll they have taken on doctors as well as patients. His response has been to set up a regular therapy group in London for fellow mental health professionals.

'This is a bit tricky, Gwen,' Ray was saying. 'I want to talk to you about a new member of my Friday group, but I don't know if there's a boundary or ethics question. She's a forensic psychiatrist, and I know it's a small world . . .' He's such a thoughtful, careful man that I wasn't surprised that he would raise this question about someone I might have had dealings with as a colleague; this was right and proper.

'Her name is Kirsty,' he added. 'She had a trainee placement at Broadmoor in the past.' I've worked with many trainees over my thirty years at the hospital, and I don't remember them all. But when Ray told me her surname, I immediately knew who she was. We had a memorable encounter once, during which I'd suggested she might find group therapy helpful. I was glad to hear that she was now working with Ray and immediately curious to know more. I told him I didn't see any conflict, given that I'd had no dealings with Kirsty for a long time and was unlikely to in the future. Our brief relationship ended when she moved on to her subsequent placement, but I remembered her like it was yesterday.

———

It was a Monday morning some five or six years earlier, and I'd just come out of a meeting with our new group of psychiatric trainees on placement at Broadmoor. They were all required to participate in peer discussions, in which they explore the emotional impact of being a psychiatrist, and I enjoyed facilitating these groups.[1] I was back at Broadmoor part-time, providing supervision and consultation to anyone who wanted it, especially trainees in forensic psychiatry.

A young woman, her pleasant face framed by a curtain of straight dark hair, was hovering in the hallway near my office. 'Dr Adshead? Could I possibly come and see you sometime about something?' She'd been in the trainee group, and I'd noticed her for her willingness to engage in the group discussion. I struggled to remember her name, then gave up.

'Of course . . . I'm so sorry, I'm at that stage of life when names escape me.'

She smiled at that. 'I'm Kirsty.'

'Of course you are. I remember now.' I waved an invitation into my room. 'Do you want to talk to me now? I have an hour before my next thing.'

Kirsty looked doubtful, already half turning away from me, as if she regretted accosting me as soon as she'd done it. 'Right now? I mean, it's not urgent . . .' But something in her manner and the strain in her face suggested it might be, at least for her. There's no time like the present. 'Why don't you come and sit down? Do you want a cup of coffee?' She nodded and followed me into the kitchen, where we made small talk about her training while I got coffee for us. My office was a rather cosy space, but I'd managed to fit in a second chair for visitors, and while she held our coffee mugs, I pulled some files off it to make room for her to sit. She didn't do so immediately; many people who came in found their eyes drawn to the two long, heavily laden bookshelves that dominated the room, and she was no different. She stood to inspect the spines, taking in a random mixture of poetry and plays (mainly Shakespeare), and an array of texts with 'madness' and 'violence' in their titles.

'Have you read all these?'

'Most of them . . . and the ones I haven't read I hope will seep into my brain by osmosis.' She laughed, and I was glad, wanting her to be at ease. I waited for her to sit, then settled opposite her. 'Tell me, what do you want to talk to me about, Kirsty? And you can call me Gwen.'

She hesitated, then blurted out her request. 'I wanted to ask you about therapy. Um, this is personal, I know, but once I qualify as a forensic psychiatrist, I'm thinking about maybe training as a therapist, like you did, but I'm not sure.' She paused, looking to me for a cue.

I remember asking one of my old trainers this same question many years earlier. It was reasonably common for psychiatrists to work as therapists back in the day, but it has become increasingly rare in the twenty-first century; I'm an odd bird among my fellows, especially in forensic psychiatry. When I was at the same stage as Kirsty in my career, I was fortunate enough to have support from people who could see the value of training psychiatrists in psychotherapy, too, so they're able to help people change their minds with therapy as well as medication.

I told her what I knew about training as a psychotherapist, warning her that things might have changed over the years. After attending years of lectures and seminars, I had to provide a set number of hours of therapy under supervision in order to qualify. In my case, that amounted to something like seven hundred hours of individual therapy and two hundred of at least two other kinds of therapy. I was also expected to be in therapy myself; it's an essential part of the training, and the minimum is usually a couple of years, although most therapists have many more. To me, it is startling that psychiatrists are not required to have any experience of receiving therapy, even though it's known that the work we do is emotionally demanding. But I'm aware that many do seek out therapy anyway, and I suspect it always makes them better at their jobs.

Kirsty was scribbling some notes on the pad she had opened on her knee, her face serious. I could see a faint pink flush creeping up her neck. This must have sounded rather daunting in terms of time commitment, and I said as much.

She gazed at the numbers and notes on her pad, a little dismayed. 'How did you do it all? Was it worth it?'

'Absolutely! It was a lot to fit in . . . but somehow, it happened. I've never regretted it. But it's a personal decision, too. You don't quite know what you'll find out about yourself, and that process isn't always comfortable. But maybe you know this already?' I needed to be careful not to assume too much.

'I . . . I'm not sure what to do.' She looked so anxious that I began to sense that she had something beyond my professional advice on her mind. It's not uncommon in my experience for people I work with to start a conversation about something professional that then strays into the personal. But there were boundaries to what I could or should hear about a trainee; sometimes you walk a line between teaching and therapy, and you can't always know where that line is in advance.

I decided to make it easier by saying she could tell me as much or as little as she liked and assuring her that we were talking confidentially – unless whatever was on her mind was affecting her professional function at the hospital. She sat there momentarily, gathering her thoughts, or possibly her courage, then raised her eyes to check in with me visually.

'The thing is, Dr Adshead, um, I went to a psychodrama workshop this weekend, and . . . well, it was difficult.' She stopped. I was intrigued – I'd done the same when I was a trainee as part of getting a sense of the different modes of therapy, and I'd found the psychodrama work very powerful. It functions rather like music on the mind; it can get under your defences and arouse deep emotions. Someone in the group will volunteer an important memory from their life, then act as the director of an improvised re-enactment, choosing people to assume the different roles. The rest of the group act as an audience and watch the 'cast' as they play out how they think the real-life scene might have gone. It is remarkable how close the re-enactment sometimes is to what occurred.

'Were you asked to improvise, Kirsty?'

'Oh no, I wasn't. I was in the audience.' She stopped short, biting her lip. I waited until she was ready to continue. 'This one woman . . . she set up a situation from when she was a teenager. Her mum had recently remarried, and she realised her younger stepbrother was secretly watching her take a shower.'

'And then what happened?'

Kirsty's head was bent, and I tilted my head slightly, non-verbally encouraging her to look up. When she did, I saw her eyes were full of emotion – was it fear?

'Well, this woman had to speak to her mother about her stepbrother, and she recalled her mother becoming angry and defensive. It was just . . . excruciating to see how that played out. So disturbing. And like I said, Dr Adshead, I wasn't even in the scene, just in the audience.' She looked down again.

'Not an easy subject,' I said carefully, not wishing to get ahead of her. She seemed to be moving into patient mode, persisting in addressing me as 'Dr Adshead', as if we were in a therapy session together, but I decided not to point that out. I thought I knew what might be on her mind.

She cleared her throat, then her words came out in a torrent. 'I just felt I had to get out of there. I was feeling scared, upset . . . I don't know. I walked out of the room and didn't go back. I felt – I still feel – so stupid. So lame. Ashamed of myself.' She paused, making a visible effort to regain control of her voice, then went on. 'I realise if I'm going to work as a forensic psychiatrist, let alone as a therapist, I'm going to have to make sure I can handle working with all kinds of offenders, without feeling negative or hostile, right? Including sexual offenders . . .' She took a breath. I held eye contact with her to show she had my complete attention.

'The thing is . . .' She paused. 'I was sexually assaulted as a child. As a teenager.'

My guess was correct.

'I'm feeling as if maybe I need to sort out my feelings about all that before I make any big decisions or figure out my next steps in training, so . . .'

I had a hunch this was the first time she had told anyone about her experience of sexual assault and asked if that was the case. A tiny nod in response. I chose my words with care. 'I'm so sorry to hear that happened to you, Kirsty. And you're wise to pay attention to disturbing emotions that arise like this and explore them further so they don't get in the way of your work, as we were discussing in the group.'

'Do you think . . . could you help me find a therapist?' She had her hands clasped so tightly that her knuckles whitened, and her lip trembled.

This was a tricky situation. There were limits to the conversation we could have, especially without any preparation, with both of us at the start of a working day. If she had never told anyone about her experience of abuse before, then how and when she did talk about it needed some planning and structure so that she would have a degree of control over any distress that might arise. Timing is so crucial in trauma work, and I felt somewhat ambushed – probably much as she had at the psychodrama workshop. I said something along these lines to her, my tone a little brisk and practical. Kirsty readily admitted she hadn't thought things through. 'It isn't just a case of coming in and saying, "Give me the name of a therapist," is it?' She smiled a little, and I thought how personable she was.

'Kirsty, you're not the first health-care professional I've met who's had such an experience and never told anyone.' I paused, considering whether I should go further. 'Could you say any more? To help us think about what type of therapy might be helpful for you?'

She sighed. 'When I chose this specialism, I was worried that it would be difficult for me to work with people who had sexually assaulted children, but then I did at my last placement, and it wasn't so bad. But now I worry about the fact that so many of the child sex offenders I've seen were abused as children and then went on to harm others. I can't help worrying if that means . . . could I also be abusive? I can't imagine it, Dr Adshead, but if I have children someday and . . .' She broke off, her distress evident.

I needed to know a little more about her experience because the context and her relationship to the abuser could make a difference to any recommendation I might make, but I didn't want this chat to stray into a therapy assessment. After a moment, I decided on what to say. 'Can I ask you one or two more questions? Would you be okay with that? You don't have to go into detail.'

She shut her eyes briefly, then took a deep breath. 'Okay. What do you want to know?'

'Was the perpetrator someone you knew?'

Her head dropped, and her voice lowered. 'It was my maths teacher. In my second year of sixth form. I think he was retired and just returned as a supply teacher to cover our usual teacher's maternity leave. Mr A, we called him. The worst thing was that I liked him at first, and I trusted him. He helped me a lot. I was *not* good at applied maths.'

'Neither was I,' I said, and was rewarded with the briefest of smiles. I wanted to reflect her words to her so she knew I'd been listening and understood. 'This man, Mr A, was someone you liked and trusted. That must have been hard.'

'It really was. He was always . . . so kind.' She was keeping her composure, but it was clearly an effort. 'Then he offered me some extra tuition, one-to-one. I didn't think anything of it; I was grateful. Once we were alone together, he grabbed me, stroking my leg, trying to kiss me, and then he took my hand and put it—' She stopped abruptly, her eyes filling with tears.

I pushed a box of tissues towards her, and she took one gratefully. 'I see this is hard to remember, Kirsty. Thank you for finding the courage to share it with me. Don't worry, you don't have to say more.' I waited for her to compose herself and offered another question only when she seemed ready. 'Were you frightened of him in the moment?'

'Frightened?' She shook her head. 'No. Not exactly. Not then. I was too shocked and embarrassed. I think I pulled away and said I had to go. I just wanted to get as far from him as possible. I turned and ran. He called after me, but I didn't look back.'

'And later?'

'I felt a bit ashamed . . . and stupid really, more than anything . . . and angry, too, and really embarrassed, you know? I didn't know what to do. I couldn't face telling anyone about it. And I knew he was leaving the school soon anyway.' She paused briefly, then added, 'The only time I felt *frightened* was on the last day of term, after his final class. Just before the bell went, he said, "Can you stay after a moment, Kirsty? I need a word." Then he started walking towards me, and I shouted, "No," really loud. Then I grabbed my things and got out of the room as fast as I could. My best friend Carol came after me, wanting to know why I'd been so rude to him, teasing me and telling me everyone thought I was his "pet". Ugh. His "pet".' Kirsty shivered at the thought. 'So horrible.'

'Horrible,' I echoed. 'It sounds like you had to be brave to manage that experience, and again, it was brave of you to share it with me today. That can't have been easy.'

Although still upset, she was curious to know why I had asked her those specific questions. We were getting back onto safer ground, the trainer and the trainee. I sat back in my chair. 'The best evidence we have suggests that the long-term outcomes for sexual abuse are less severe if the abuse is a brief, one-off experience with someone who is not emotionally important to the victim. This also tends to be the case if people are more ashamed than frightened, because the stress of shame can be easier to bear than high levels of fear.'[2]

'Really?' She frowned, and I noticed her hands were now clenched tightly in her lap. I could see she was struggling to process what I was telling her, which didn't surprise me. Despite evidence to the contrary, it seems hard for people to accept that minds change all the time and it is possible to survive difficult events, and to survive them well. Everything I have learned in my years as a trauma therapist has shown me that not all bad experiences permanently wound us. So much depends on the context and the meaning we make of them at the time, and of course, any previous experiences of wounding and their consequences.

This is especially true of sexual abuse that occurs when the young mind is still forming. There is a tendency to see this as 'the worst kind' of trauma, and one that will stay trapped in the mind forever. The truth is more nuanced, because it's hard to separate out the effect of sexual abuse from other kinds: these children are often also exposed to physical harm, emotional maltreatment and neglect, which makes parsing out the effect of sexual abuse difficult.

In addition, sexual abuse tends to happen after the age of ten, especially to girls entering puberty and adolescence. The perpetrator is unlikely to be a stranger in an alley; the most common scenario is that they are assaulted by one of their peers, very often an underage boyfriend.[3] Sadly, children who suffer any form of maltreatment are at increased risk of victimisation in adulthood, which again makes working out the specific impact of their childhood experience especially complex. I had only to think of people like the Irish siblings I had met, Mick and Bridie, whose lives took such different courses after their similarly traumatic childhood experiences.

'Kirsty, I think that experience will not make you risky to others. But I support your intuition that it could affect how you react to treating or assessing people who assault others by stealth or exploit their trust. We will meet such people in our work – both men and women. Being prepared for that is an act of self-care and resilience.'

'You don't think it makes me . . .' – she searched for the right word – '. . . damaged? Even . . . unsuited to this job?'

A teachable moment in so many ways. 'Kirsty, if you were a therapist right now and someone asked you that question, how would you respond?'

'Do therapists always answer questions with questions?'

We both laughed at that, and I was pleased to see she'd unclenched her hands and her shoulders were no longer up around her ears. Maybe she felt a bit lighter after getting this story out. How long had it been, a dozen years or more? A long march with a heavy burden.

She hazarded an answer to my query. 'I think I'd tell the patient that, um . . . experiences like this are upsetting, but . . . they don't have to . . .'

She was floundering, so I picked up the thread. 'But upset is not the same as damage. In fact, it can be an indicator of good mental health if you feel bad when bad things are happening.' Time was getting on, and I needed to close this conversation safely. 'How do you feel now, Kirsty? Are you okay to work the rest of the day, or would you like me to have a word with your consultant? I'm sure they wouldn't mind if you needed to take—'

'No!' Kirsty was vehement. 'I'm fine. I don't want anyone to know anything. Please.'

'What you've said here won't leave this room, Kirsty. But there's no shame in taking time for yourself if you're feeling vulnerable.'

She shook her head. 'I'm okay. And I'm sorry, Dr Adshead. I didn't think this would turn into more of a therapy discussion than a career chat . . .'

'It happens that way sometimes,' I said. 'I'll email you some websites with registers of trained therapists. In my experience, people often need to meet a few different ones to find the right person, and you might want to think about whether you prefer a male or female therapist. You could also consider group therapy, not just individual.'

Kirsty blanched. 'Groups? With other people? I never thought of doing that . . . for this.' I wondered if her experience in the psychodrama workshop had coloured her view of other kinds of group work.

'I get it,' I told her. 'But as a group therapist myself, and as someone who's been a patient in group therapy, I can tell you that in groups, you learn things that you can't learn anywhere else.'

'I'll keep that in mind.' She sounded unconvinced.

'Oh, and look for someone who works near where you live, if possible,' I told her. 'That way, you can schedule sessions around going to and from work. I remember my first therapist lived around the corner from my flat, so it was easy to arrange to meet.'

Kirsty stared at me. 'Your *first* therapist? How many have you had?'

'Three,' I said promptly. 'The first to learn how to think about my mind at all; the second to train me to be a group therapist and learn about other minds; and the third to keep me from going mad when I was in trouble.'

'That's a lot!'

I laughed. 'Well, we need different kinds of help at different times in our lives – and I imagine the time may come when I will need it again. Who knows?'

I stood to indicate that our meeting was ending, and she thanked me again. I watched her as she headed down the corridor, turning back to wave as she rounded the corner.

I thought she would make an excellent consultant forensic psychiatrist, and I hoped she would get some therapy and train as a therapist, too. She didn't come back and talk to me individually again, and when she left Broadmoor at the end of her placement, I didn't expect to cross paths with her in future – and in the literal sense, I never did. However, years later, through the synchronicity of my connection with Ray, I would learn more about her story, gaining another chance to marvel at how human minds can change so profoundly when people are willing to risk exploring tricky territory.

———

I did not speak about that past encounter on my Zoom call with Ray. As I had promised Kirsty, it was between us. Anyway, I was there to hear about *his* experience with Kirsty and work out what was concerning him.

He told me they had begun working one to one, although he mentioned his group to her right away, explaining that people often join one after some individual preparation with the therapist. When he began taking her history (of which I knew nothing), Ray learned she had grown up in the Surrey suburbs, the only child of two successful professionals who were keen for her to achieve. She had done well at school and had always wanted to be a doctor. 'What else?' Ray muttered, pages rustling off-screen as he went through his clinical notes. 'Here you go. Then she said exactly this' – he made air quotes – '"I had a great childhood."' He asked her if she could talk a little about how things were at home, and in response, she volunteered that she was 'always closer to her father than her mother', but she didn't elaborate, even when he tried to encourage her to go into more detail.

'She blocked me, Gwen.'

'She blocked you?'

'Just shrugged off my questions, waving her hand and saying something like, "It was all a long time ago." What's that about?'

I could see that Ray was frustrated, and we talked a bit about those brief, bland phrases that people sometimes use in therapy, things like 'I can't say' or 'I don't recall' – all classic linguistic tricks that I now associate with someone with avoidant attachment, who shuts down discussions of distress and uses neutral phrasing as a way to deactivate feelings.

In subsequent therapy sessions, they inched a little deeper, but Kirsty still spoke in generalities about things like 'nice family holidays' when she was a child and cousins they would visit by the seaside. Ray admitted that he felt she was emphasising the 'ordinariness' of it all. To him, she seemed deliberately over-cheerful about her Great Childhood. But once they had built some trust, she revealed her assault experience with her teacher, and he felt they were finally getting somewhere. I did, too, although I did not react to that revelation. Again, this conversation was about Ray's response to Kirsty, not mine.

As she described the encounter with Mr A to Ray, Kirsty became emotional for the first time, breaking down and sobbing in front of him, quite the opposite of her calm affect and approach when speaking about her early years. I was intrigued when Ray added that Kirsty confided to him that this traumatic memory was so painful that she had never shared the story with anyone before. I wasn't sure what to make of that. I didn't think it was a deliberate lie, but had she blocked our long-ago conversation from her mind? Or perhaps she found it a bit embarrassing to admit to Ray that she'd spoken to her trainer about it at one time and didn't want to bring me into their conversation? I kept these questions to myself.

Ray and Kirsty had gone on meeting one-on-one for a while, mainly talking about Mr A and her feelings about how this past trauma was affecting her work now. For example, she was doing more assessment work for the family court, and she told him she was finding it difficult to deal with any case involving possible child sexual abuse. Eventually, Ray reintroduced the idea that she might join his therapy group, alongside other medical professionals grappling with mental health difficulties. She was willing to try, she told him, and he took that as a good sign of progress.

'But now I'm not so sure . . .' He stopped.

'What's the problem?' I tried not to speculate.

'Look, she's a good listener and thoughtful about her issues, but . . . I'm not sure the group is working out.' At first, he explained, Kirsty had seemed open in the sessions; she'd even been able to talk with the others about the sexual assault by Mr A. They were warm and supportive, and their response seemed to enable her to reveal more, including some other incidents of coercion and abuse involving a couple of recent boyfriends, stories that Ray hadn't heard before.

'That sounds constructive. What's your worry?'

'I don't know . . . It's odd. She's getting into that over-cheerfulness I saw before, where she'll just blank us.' Could he give me an example? He scanned his notes again. 'Okay. One day, we start discussing attachment styles, and someone in the group asks Kirsty which carer she was close to as a child. She mentions her dad, as she had done with me. Then she was asked for a specific memory of the closeness, of course.' He looked at me in acknowledgement that we'd often talked together about this line of questioning, drawn from the Adult Attachment Interview protocol, which we both knew well. 'Here we go, I thought. She's going to block this for sure . . . But she didn't. At least . . . well, let me talk you through this properly.' He turned back to his notes.

Kirsty had offered up a memory from when she was eight or nine. As a special birthday treat, her father had taken her to London to visit Liberty, the grand department store. She was 'over the moon' when he told her to choose her birthday dress, anything she liked. Ray read, '"I picked one with lovely swirly floral patterns and red velvet trimmings."' Kirsty said she loved this dress and wore it whenever she and her father went on outings together, usually to concerts. He was musical, and so was she; this shared interest added to their closeness.

Ray stopped. 'Good answer, as far as it goes.'

I asked the obvious question: 'No mention of her mother?'

'Exactly!' said Ray. 'When someone asked if her mother had joined them on any of these outings, she just looked blank and trotted out that line again: "It was all a long time ago." When pushed, she became cold, almost irritated, saying, "Sorry, I can't remember if she did or not." It was strange.'

'And then?' I asked.

'How can I describe this, Gwen? It's like she pressed the mute button and went dumb. She just slumped as if exhausted. Shut down the conversation.'

A line from the Psalms popped into my consciousness: 'I am so troubled that I cannot speak.'[4]

'How did the group react?'

'I could see they didn't like it. Tina – you know, the nurse I've mentioned to you before – she wasn't having it and blurted out something like, "Wait. You must remember *something*."'

'And then?'

'Kirsty turned to look at her, and I thought she might be about to push back. Instead, she just repeated, "Sorry, Tina. It was a long time ago."'

This was puzzling. 'What did you think was going on, Ray?'

'To be honest, I hadn't a clue. I could see she was blocking, and the other group members could, too . . . but I felt like I was paralysed. Then we moved on to something else, but I could tell the group wasn't happy. Including Kirsty.'

I told him I might have done the same; we let the group do the work. Now, I wanted to pick up on something in Ray's language.

'You said you were paralysed? What was that about, do you think?' I had an idea, but I wanted to hear him out first.

He blew out a heavy sigh. 'Well, I kept thinking about the maths teacher, and I didn't want to step in and be, you know, an intrusive male authority figure making her do something she didn't want . . . but maybe I should have intervened?'

'If there's something important there, Ray, it'll come back around . . . You know how that goes.'

'Fair point,' he said. 'I do realise I don't "have to" do anything. But now I'm feeling so frustrated. Angry, even. Not with her exactly, but . . .' He trailed off, lost in thought.

After a minute, I observed, 'Now you're silenced, too.'

He didn't reply. Knowing him as I do, I could sense how hard he was trying to work out what he needed to articulate. 'Okay,' he said after a bit. 'So, there's a quality to her speechlessness that unnerves me. Shuts me up.'

'Like, it's bad to go there?'

'Maybe.' Ray glanced at his watch. We were overrunning a little, which I try not to do, but sometimes it's necessary in a supervision. 'Thanks for this, Gwen. Ironically, it's been helpful to talk to you about how my patient can't talk!'

Kirsty could talk quite fluently, I pointed out. Remember how she'd told him about her maths teacher? That was something. Maybe she'd get to the point where she could say more about her parents. As both a patient and a therapist, I knew it took time to find your voice in the group and dismantle those walls of language and silence that we build for our protection.

When therapists hit a 'block' like Kirsty's, exercising patience is essential, but I did feel for Ray. Humans use silence in different ways in different spaces, and you don't have to be trained as a psychotherapist to realise that some kinds of silence are more difficult to manage than others. They come up every day: the respectful silence in a house of worship or library; an angry dumbness that any parent of a teen will recognise; the wordlessness of an uncomprehending student; the shocked silence after sudden violence or an unexpected loss – the variations are infinite. It is only natural in the face of baffling silences that we humans feel the urge to deploy our unique advantage of language to gain understanding, and we can resent being thwarted in that effort.

As we ended our session, I told Ray about a remarkable talk I'd read about, which the American poet and activist Audre Lorde had given back in the 1970s, after she was diagnosed with a cancer that could have been fatal. She described her experience of fear of death, and how she'd found strength and hope in coming together with other women. I loved her testament to the power of a group in aiding healing and her passionate belief in the power of the 'transformation of silence into language and action'.

'Hang on a minute.' I reached up to the shelves by my desk and extracted the well-worn text, thumbing through the pages quickly. Then I shared with him this quote, which Lorde attributes to her daughter: '. . . you're never really a whole person if you remain silent because there's always that one little piece inside you that wants to be spoken out, and if you keep ignoring it, it gets madder and madder and hotter and hotter, and if you don't speak it out, one day it will just up and punch you in the mouth from the inside.'[5]

———

The next time Ray came online for our call, he was unmuted and keen to share his news with me. 'I think Kirsty's at boiling point,' he told me.

A new member had joined his group, an older woman called Jan, who was nearing retirement as a surgeon and stepping down from high-profile roles in medical management and administration. Ray reported that Jan and Kirsty had taken an instant dislike to each other. 'I could see Kirsty was trying to be nice at first, complimenting Jan and fluttering around, offering her a chair, that sort of thing. But Jan is very no-nonsense, old-school, quite formal. Something about Kirsty's manner seemed to set her off, and she was a bit hard on her.'

'How so?'

He explained that as soon as Kirsty offered any comment, Jan would jump in and disagree, almost on principle. She wasn't like this with anyone else, and the others picked up on it. After a couple of sessions like this, Ray encouraged the group to talk about it openly, and then Jan came right out and said she found Kirsty to be false and superficial, 'a classic people-pleaser'. Kirsty, who was generally quite assertive and dynamic in the group, didn't defend herself but just sat there looking miserable, as Ray described it. He knew he had to address the conflict but needed to figure out how. And then, as luck would have it, Jan couldn't make it to the next week's session.

'I started by asking Kirsty how she was feeling about the group – not mentioning Jan, but just in general.' Right away, she stated she wasn't getting anything from it any more and thought she should quit. 'Then Tina jumped in and asked her, point-blank, "Is this because of Jan?"'

I winced a little, and Ray groaned. 'I know. So truthful, but so awkward. Then Kirsty didn't answer. There was just silence – a tense silence. I let it run for what seemed like ages. And then she started crying. Quiet, choking sobs, her shoulders shaking. Somehow so much worse than a noisy flood of tears, you know?'

I leaned towards the screen, chin resting on my hand, intrigued.

Ray went on: 'Once she settled, I asked her if she wanted to say anything. She shook her head no, so I opened things up to the group. "What do we think about what's happening here between

Jan and Kirsty?" Then, before anyone could offer an opinion, Kirsty said something mumbly, her voice so low that I had difficulty understanding. I asked her if she could repeat it. Wait till you hear this, Gwen.'

'What did she say?'

Ray read his note out to me verbatim: "'I'm scared of Jan. She's just like Mum. She hates me.'" He set down his notes. 'How about that?'

It was my turn to be wordless while I tried to make sense of it. Eventually, I was able to ask Ray how he'd felt in the moment.

He frowned. 'I don't recall any strong emotion, apart from concern. The other group members tried to connect with her, responding to what she'd said, but she just shook them off. And then she started repeating, like an incantation or something, "It was all a long time ago, it was all a long time ago . . ."'

That sounded unnerving. 'What was *that* like for you?'

'It was weird. It seemed like she was trying to soothe herself,' said Ray. 'But it wasn't working very well. She was shaking uncontrollably, and then she started hugging herself and rocking back and forth . . . That got me worried.'

I had a thought: Kirsty's language, especially her use of the present tense, indicated that she might be experiencing her past relationship with her mum in the here and now through her interaction with Jan. Her fear was very 'live', even if she was in no danger, which indicated a disconnect with reality. I said as much to Ray.

'Yeah. I thought she was dissociating. It was truly alarming, Gwen.'

Dissociation is a process by which people alter their consciousness, effectively 'tuning out' from their current situation. Everyone's conscious awareness fluctuates at times, like when your mind wanders while driving during long car journeys, but there's a more extreme alteration of consciousness that trauma-tised people sometimes employ when they feel overwhelmed by distress: they step outside of what is happening in real time, which makes them hard to reach. It's a way of coping that children use, as I'd seen with IRB claimants such as Mick and Bridie, but most adults do it only if they feel overwhelmed by pain. It can happen unpredictably with people in therapy if they encounter a scenario that reminds them of something painful in their past, giving rise to what is known as a 'situational memory'.[6] What past event was coming up for Kirsty in the group?

'Did she say anything else, Ray?'

'I think she was in touch with being a child,' said Ray, 'because the next thing she said was . . . Hang on.' He consulted his notes. '"I must be horrid because Mummy doesn't like me." And then she repeated over and over, "There's nothing I can do, nothing I can do."'

'Wow. Something that's hard to hear, Ray. What did everyone do?'

'Well, we tried to reach her, but she was shaking her head back and forth, resisting any approach, so we stopped. Unfortunately, it was now near the end of the session, so I said something like, "Kirsty, the whole group can see your distress and how much pain you are carrying about this relationship. Maybe you can't say much more right now, but we're here for you and ready to listen, when you feel ready. Maybe when we meet next time?"'

'I like that, Ray,' I said, mentally filing his excellent words away to use myself sometime, if needed. 'You brought everyone in, end-ing not just on Kirsty but the group as a whole. What happened next?'

Ray explained that as the session ended, he went and sat next to Kirsty and checked if she felt able to go home safely. He noted that her body movements had ceased, and she seemed to have reverted to her usual self, apologising and assuring him she was okay. She appeared eager to get out the door, and he had to leave it there.

———

When the next group session started, Ray was relieved when Kirsty turned up as usual. Everyone was present, including Jan, who was back from holiday. Right away, Jan surveyed the room and picked up on some change in the dynamic. She wasn't one to mince words. 'Well? What did I miss?'

'I could tell the group wasn't sure how to respond, given what Kirsty had said about Jan while she was absent,' said Ray. He thought they were probably all worried about upsetting Kirsty again. Jan got a bit impatient, demanding an update, as anyone might if they felt left out of some inside story. Eventually, Tina spoke up, forthright as ever: 'You upset Kirsty. She was in bits last week.' Ray said Jan didn't miss a beat but turned immediately to Kirsty, asking, 'Did I? How?' Ray told me he was holding his breath, waiting to see how she would respond and hoping the others could let this play out. I was right there with him, on tenterhooks.

Kirsty maintained her calm, turning to Jan and confirming that she had upset her. From the first time they met, Kirsty said, she had tried to be friendly, and in return, Jan had only been unkind to her for no reason. Cruel, even.

'"Cruel",' I repeated. 'Wow. How did Jan deal with that?'

'She looked pretty stunned,' Ray said. 'Didn't even try to defend herself. Kirsty then told us that it was only in Jan's absence that she'd realised how much Jan reminded her of her mother, whom she could never please. Then she looked at Jan and said . . .' – he glanced at his notes to get the words right – '. . . "I can't bear to be in the same room with you any more."'

'Wow,' I said again. This was quite an exchange. 'Then what?'

Jan had appealed to Ray, ashen-faced: 'What do we do now?'

I thought I knew what Ray would do, because there was only one thing *to* do as the group's conductor: if possible, he had to guide them towards learning more about Kirsty's experience and understanding how her past relationship with her mother came to be embodied in her present relationship with Jan. I was glad when he confirmed that was exactly how things had gone. 'And then it all came out,' he told me. 'About her mother. I've heard a lot in my time, Gwen, but this was grim. So sad.'

He took off his glasses momentarily, rubbing his eyes.

'What on earth did this mother do?' I felt wary of what I was going to hear.

So much for Kirsty's Great Childhood. The group learned that her early years had been fraught with fear and humiliation, a catalogue of chronic physical abuse and intimidation by her mother, all happening behind the walls of her ordinary, comfortable suburban home. Ray was of the view that the mother must have been mentally unwell, though Kirsty had not said as much.

'A couple of the incidents she spoke about have stayed with me, and I need to think them through with you today, Gwen,' Ray said. One of the primary purposes of supervision is to work on a therapist's responses to complex material. In the therapy room, the focus must be on the patient's emotions and experience, but here on Zoom, it was Ray's turn. I told him I was ready to hear whatever he wanted to tell me.

'The first thing was shocking. Kirsty said she was about ten, preparing for an entrance exam to a posh secondary school that her mother had selected. She did some mock tests, which didn't go very well, and when her mum found out, she went berserk. Apparently, she dragged her upstairs to her room, threw her on the bed and then went to the cabinet Kirsty used to display a collection of little china ornaments and dolls. She started smashing them, one by one, until they were all destroyed.'

What a bizarre vignette. And it got worse.

'Then the mother grabbed Kirsty, shouting and slapping her, pushing her up against a wall, pinning her arms . . . and then . . . Gwen, she *bit* her.' He shook his head, anger and dismay evident in his face.

In the square on the screen next to him, I saw my own reaction: I gasped, and involuntarily, my hand flew up to cover my mouth. This is a common response to seeing or hearing something upsetting, and I'm not sure why, except it usually means we don't want to let something out that we can't control. I felt like I was stopping a noise, stifling a scream.

After a moment, I mustered the only question I could: '*She bit her?* How? Where?'

'On her arm, I think . . . or maybe she said her hand? Someone in the group asked, but I was too focused on Kirsty to take in all the details,' Ray said. 'Imagine this grown woman, this nice Home Counties mum, holding her child against a wall and *biting* her . . . like an animal.' He grimaced. 'An untrained dog.'

'What happened then?'

'Kirsty told us she couldn't cry out for help or do anything to stop her mother. Her exact words were, "I was speechless with terror." Then she said she felt ashamed that she'd disappointed her mother and made her angry.

'Oh my word, Ray . . . What were you feeling at this point? I'm aware of feeling stunned just listening to you.'

'Same. And I could see the others were, too. When Kirsty stopped speaking, you could have heard the proverbial pin drop – for once in this group, there was nothing to say.'

He raked his hands through his hair so that it stood on end, giving him an Einstein-ish air. 'And then, in the next session, she told us this other awful story, something that happened maybe a year or so later.'

Kirsty had invited her best friend Linda for a sleepover. At some point, her mother had walked in on them play-fighting, rolling around on the floor together and laughing. Without explanation, she grabbed Linda by the arm and pulled her out of the room, then called her parents to come and collect her immediately. Kirsty was utterly mystified – she couldn't think what she'd done wrong this time, but dreaded finding out. She climbed under her bedclothes, waiting for the worst. Minutes later, her mother marched in with kitchen scissors, yanked Kirsty's favourite dress out of the closet – the one her father had bought at Liberty – and cut it to pieces, throwing the scraps at the cowering child and telling her, 'There. That's what you get for being a lesbian.'

I was conjuring the ugly scene in my mind's eye: the terrified little girl huddled on her bed, covered in shards of fabric like so much shrapnel, no doubt fearing physical violence again and feeling utterly shamed by her mother's contempt and rejection. It was so tragic. So sad. It occurred to me that these experiences with her mother might have influenced Kirsty's choice of specialism: she would make a career out of trying to understand and care for other people's unpredictable rage and cruelty.

Ray watched me process what he'd said, waiting for my response. With some effort, I brought myself back into the room.

'So . . . she's confused Jan with this mother she could never please? Who was so cruel?'

'Exactly where I went, Gwen. I'm thinking Jan's hostility activated Kirsty's experience of fear and vulnerability in a way she could no longer suppress. Jan wasn't so bad, just a bit off with her, but maybe because we'd talked about that incident with her teacher, Kirsty was already feeling vulnerable in the group and, therefore, more raw, more open to being hurt?'

This made total sense to me. I have seen over many years how the process of therapy can gradually disclose layers of experience and identity as the unspoken becomes articulated and heard. Like archaeologists, therapists try not to dig too deep, too quickly, instead gently brushing what comes to the surface, until, layer by layer, a full outline of people's distress emerges.

The remarkable way in which Kirsty's story emerged in Ray's group reminded me that traumatic stress might be expressed differently if it's been hidden for a while. I considered whether it was harder to put distress into words after a long time has elapsed, especially if the feelings accompanying it are rage and shame. It's not easy to research why people choose not to speak, especially if they think it has kept them safe in some way.

Kirsty's case also made me question whether, when the words do come, people discover that experiences of fear and vulnerability seem more acceptable in our culture. By contrast, there's a common anxiety that angry voices aren't welcome, especially if they are female. A powerful recent paper calls for a commitment to address the 'gender gap' in trauma therapy, stating that women can be in a double bind after trauma. This study found that clinicians often do not even ask female patients about their anger, or if it comes up, it is treated as a symptom of something else, such as depression or anxiety.[7] Sending them to anger management groups, which may be made up of angry men, can be unhelpful. Perhaps new, gender-specific strategies are needed.

Thinking of Kirsty's experience also reminded me of how early attachments influence our capacity to speak about experience, both remembered and in the present. Some people with attachment insecurity secondary to abuse can struggle to tell their own stories; in technical language, they have 'decreased autobiographical competence'. They may also struggle to find the right words for painful feelings.[8] This seemed to describe Kirsty perfectly.

The point about therapy is that it offers a platform to stand on and find your voice. We talk a lot about the 'secure base' we need in our early life, which we might also develop later by working with a good therapist or, as in Bridie's case, making secure adult attachments outside of therapy. But if we think about therapy and supervision as a 'secure embrace' for both parties, a container in which you can leave fragile things to be looked after and safely held, then this can be where group therapy comes into its own. Yes, meeting Jan had ignited negative memories in Kirsty; still, I suspected that making good attachments to both Ray and the group had enabled the traumatic events she'd suppressed for so long to crystallise into conscious language and allowed her the courage to speak.

Though I could not get into this with Ray, I thought about that meeting with Kirsty in my office years before, when she first spoke about her experience of sexual abuse. She had been prompted by that workshop scenario involving a young woman confronting her mother about something painful and uncomfortable. She had expressed a worry then that having been sexually abused, she might become an abuser when she had children. With her new disclosure in Ray's group, that now made more sense to me: underneath her professional concern was a terror that she might be like her mother.

Why had the sexual abuse in her memory been more accessible, more speakable, first to me and then to Ray, than her emotional and physical abuse at the hands of her mother? With me, I thought it might have seemed a more acceptable and recognisable explanation for her response to the psychodrama workshop, one that did not require her to go deeper. Humans crave narratives that contain and explain our fear, and our memories help us to do this, so the threat is reduced.

I couldn't help but think that it is somehow more acceptable, even expected, for women to talk about sexual victimisation by men than it is to discuss cruelty at the hands of other women, especially those who give birth to us. Does sexual abuse, particularly in childhood, also function as a psychological trump card, which can allow people to access help for other traumas? Or might the attention it demands mask another, possibly even more disturbing past distress? If that's true, some intense experiences of shame, rage and grievance may suffocate under a blanket of passivity and helplessness and become unspeakable.

Ray and I moved on to musing about Kirsty's father, curious about what part he'd played. Had he known? Had he looked the other way?

'She's said very little about that, at least so far,' said Ray. 'All she's ever told us is that she was closer to him than to her mother, which, as it turns out, isn't saying much. Now I'm thinking, what if she experienced me in the group as the father who didn't protect her?'

Our time was nearly up, and we agreed to pick this up in our next session. Just listening to this story had exhausted me. Not for the first time, it had brought home the challenge psychotherapists face in carrying so much pain and anger, especially in group work, where therapists hold several minds in mind at once. Thankfully, we have supervision and peer support groups, and Ray reiterated how grateful he was to be able to bring this story to me.

'I just needed to leave her pain somewhere else, you know?'

The secure embrace. I did know.

'The good thing is,' Ray said, 'I think this has been a bit of a breakthrough for both women. Not comfortable, of course, and it's early days, but since all this came out, Jan has been able to apologise to Kirsty in her rather gruff way, which I think was a healing thing. She also said that she realised she did have some hostility towards Kirsty because of her youth and that she might envy her for being at the start of her career, while Jan was facing imminent retirement. Then people started talking about how hard it is for doctors to be vulnerable, and that brought out Jan's fear that she's not sure who she'll be when she's no longer working as a surgeon. Her work has defined her for so long, she says, and in the future, she's afraid she won't be heard, she'll lose her voice.'

I knew about this, too.

———

In later sessions, I was glad to hear from Ray that the relationship between these two women had continued to change for the better. As I think about Kirsty's experience now, it seems to me that her choice to become a forensic psychiatrist might have been a rather sensible psychological solution to her childhood distress and fear. I believe everyone who is fortunate enough to be able to choose their livelihood, as I was, will choose work that speaks to something in them that has meaning, and that meaning is different for each of us. Forensic work, like teaching, can be vocational; one of my mentors started as a GP and worked with an NHS team serving a high-security prison, an exposure that humanised violent offenders for him and 'hooked him' into this specialism. I don't have any data on this, but anecdotally, I know of other forensic professionals whose career choice was influenced by being raised in circumstances where there wasn't a lot of control, and where, in some instances, like Kirsty's, they were exposed to violence as children. I would be interested to know how many forensic professionals may have been frightened children once and found that their professional identity allowed them a helpful sense of control and mastery over scary and unpredictable adults.

––––––

Not long ago, I was asked to work on a case for the family court. I arranged to see a young woman who was struggling to care for her children and who had been accused of abuse and neglect. She'd had expert assessments before, and in reading those reports, I found one by Dr Kirsty T, prepared a year or so earlier. It was the same Kirsty, who was now studying and teaching about women and violence, as I have done. I was moved by her eloquent analysis of the young woman's state of mind and the problems she had faced in becoming a mother. She discussed how this woman had been carrying some unresolved pain and then inflicted it on her children when she became a mother, referencing theories of cross-generational trauma. It concluded with something to the effect of 'If this mother had been able to get help while pregnant, or better still, earlier in life, she might not have transmitted her pain.'

Kirsty's report revealed a deep psychological understanding and compassion for a woman not unlike her own mother. It suggested to me that she'd taken up a new perspective on her abuse. I hoped she might now see that if her mother had been given the mental health support she'd needed, she might not have lashed out at Kirsty years ago or inflicted her own madness and cruelty on her child. It was plain how much work Kirsty had done on her mind during group therapy with Ray; she had been able to think empathically about herself as a speechless, shaking child, hurt and shamed by her mother's rejection. The interaction with the other members seemed to have allowed her to leave that childhood identity behind and pursue her work as a woman of courage, meeting and helping offenders with empathy and objectivity.

I've written and spoken so much about what I've learned from my patients, but I feel equally privileged to work alongside thoughtful colleagues who continue to school me in the miraculous capacity of the mind to change – people like Kirsty, Ray and so many more. I also give thanks to the many poets, playwrights and scholars whose works line my shelves and continue to illuminate every crevice and corner of the human mind. I often return to Audre Lorde's wisdom and profound reflections on silence and language, which resonate through the years and in the case stories I've described in this book. I think she speaks to, and for, all survivors of fear and trauma when she says, 'What is most important to me must be spoken, made verbal and shared, even at the risk of having it bruised or misunderstood,' warning other survivors that 'Your silence will not protect you.' Describing the trauma of her cancer scare and fear of imminent death, she theorised that we have been socialised to respect fear more than our needs for language and definition. 'While we wait in silence for that final luxury of fearlessness, the weight of that silence will choke us.'

As she wisely concludes: 'It is not difference which immobilises us, but silence. And there are so many silences to be broken.'[9]

Coda

I end this book with many questions that have yet to be answered; in fact, they may be unanswerable. Is it 'bad' for people to decide not to talk about their past? What is the difference between those who unconsciously suppress their memories and those who do so deliberately? And rather than burying their distress, do those who speak up earlier do better, and what does 'better' look like? What difference does it make if people are able to access prompt treatment, and can the expectation of recovery aid its progress?

And crucially, why have our mental health services been so degraded? It is a terrible irony that we've never known more about effective psychological therapies and interventions for trauma than we do today – so much more than we did when I started out – yet our mental health services have been decimated by funding cuts, and people with traumatic stress reactions of any sort are likely to be turned away from the services they require.

How did we get here? Less than twenty years ago, NHS psychiatrists would provide long-term support to people, and psychological therapies were offered to those who needed them. From my perspective as a clinician, all the public discourse and media attention given today to trauma and the need for better mental health care is just talk until substantial resources are allocated that can provide adequate services to those who are suffering. We spend millions of pounds of public money on investigations and commissions that establish that childhood adversity is endemic, for example, yet a small fraction of that is allocated to mental health service providers.

I am hearing from young psychiatrists today who say that they are instructed to tell people, 'If your problem needs something other than medication, then we can't offer treatment at all.' How is it that we have allowed mental health to be less of a priority than physical health? I cannot help but think that in the future, my psychiatric descendants will be astounded at our failure to care and to act, much as we now condemn past societies and professionals for misunderstanding and misdiagnosing indications of serious mental distress.

Public health policy must change so that we treat mental suffering as we do physical suffering. Isn't a society only as resilient and capable as the mental health of its citizens? The lack of services is a concern not just for trauma therapists, researchers and policy-makers, but for everyone, because across a long and happy life, it is more than likely that we will all, at some point, endure something that will damage our sense of self and even break or destroy our identity. Rebuilding after trauma takes time, but it is achievable with a range of proven interventions. Even then, a very small number of people will find it hard to discover a new way of living after trauma, and we urgently need to understand who they are and how researchers and clinicians might be able to help.

Everyone has their wounds and their own stories to tell. I am aware that reading this book may activate all sorts of feelings in some readers, and I invite people to take care of themselves. If possible, seek out a listener and 'give sorrow words',[1] sharing it with someone who may be able to help you towards healing, change and peace of mind.

Transformation like this involves deep work of the mind and heart, and what Viktor Frankl called the greatest courage of all: the courage to suffer. I am filled with awe and respect for all the people I've seen who have been willing to walk that hard road. It has become a cliché to talk about *kintsugi* in relation to trauma, the Japanese art of piecing together shattered pottery with liquid gold, giving the repaired whole a different sort of beauty. But a cliché is only a truth oft-repeated, and an attempt to communicate something that has meaning to many. I believe the 'gold' that binds together this collection of stories is made of my memory of the bravery shown by the many people who have trusted me with their sorrow and pain.

I hope I have done them justice. These are men and women who knew great fear yet took a risk: they spoke of the unspeakable to me, a stranger, and thus found a way to not so much get over as get through dreadful experiences. Over time, I saw many of them turn terrible memories from something destructive into something creative. They survived not by recovering their old self but by discovering a new one.

Acknowledgements and References

Our gratitude to our remarkable and supportive adult children, Dan and Jack Ferris and Lily Slater.

And a heartfelt thank you to our dream team, Sophie Lambert and Laura Hassan, as well as Hannah Knowles for her early input.

For each chapter, I acknowledge below colleagues to whom I am indebted, many of whom are not mentioned in the text. But first, a few recommendations for further reading.

There are so many textbooks on traumatic stress responses and PTSD. I found these ones particularly helpful:

Figley, C. R. (Ed.), *Trauma and Its Wake, Vol. I: The Study and Treatment of Post-Traumatic Stress Disorder* (New York: Raven Press, 1985).

Van der Kolk, B., *The Body Keeps the Score: Brain, Mind, and Body in the Healing of Trauma* (New York: Viking, 2014, and many subsequent editions).

On the study of attachment theory, John Bowlby's original trilogy is excellent, but for the general reader, his introductory book (based on a series of lectures) is a great place to start: Bowlby, J., *A Secure Base: Parent–Child Attachment and Healthy Human Development* (London: Routledge, 1988).

I also highly recommend Peter Fonagy's work, including his early book with Dr Jon G. Allen and Professor Anthony Bateman, *Mentalizing in Clinical Practice* (Arlington, VA: American Psychiatric Publishing, 2008).

Professor Scott Henderson, an early adopter of Bowlby's ideas, wrote a very good paper on attachment: 'Care-Eliciting Behavior in Man', *Journal of Nervous and Mental Disease*, 159:3 (1974), pp. 172–81.

The POW

Thanks to Deborah Tomkinson for her memories of her father's experiences as a POW in the Second World War. Thanks also to many different colleagues in military psychiatric services.

Haley, S. A., 'When the Patient Reports Atrocities: Specific Treatment Considerations of the Vietnam Veteran', *Archives of General Psychiatry*, 30:2 (1974), pp. 191–6.

Shay, J., 'Learning About Combat Stress from Homer's *Iliad*', *Journal of Traumatic Stress*, 4:4 (1991), pp. 561–79.

———— *Achilles in Vietnam: Combat Trauma and the Undoing of Character* (New York: Simon and Schuster, 1995).

Shephard, B., *A War of Nerves: Soldiers and Psychiatrists in the Twentieth Century* (Boston: Harvard University Press, 2001).

For those interested in reading more about the Japanese POW experience and its consequences, I thoroughly recommend Richard Flanagan's two books, *The Narrow Road to the Deep North* (Chatto & Windus: London, 2014) and *Question 7* (Vintage: London, 2024).

The Refugee

Thanks to the therapists working in the trauma clinic at the Middlesex Hospital, and especially to Dr Stuart Turner, Dr Peter Scragg and Dr Deborah Lee. Thanks also to the volunteers at the Listening Place for discussions about traumatic stress in people who have migrated to the UK in great fear and distress and have had to survive.

Clare, G., *Last Waltz in Vienna* (New York: Holt Rinehart and Winston, 1982).

Gilbert, M., *The Boys: The Story of 732 Young Concentration Camp Survivors* (New York: Henry Holt & Co., 1997).

The Lucky Ones

Thanks to Dr Sameer Sarkar and Dr Richard Latham for discussions about civil law cases. Special thanks to my colleagues in the Perinatal Psychiatry teaching group, especially Dr Liz Macdonald Clifford, Dr Lucinda Green and Dr Maddalena Miele.

For a useful overview, see:

Baldaçara, L., Da Silveira Leite, V., Silva Teles, A. L. and Da Silva, A. G., 'Puerperal Psychosis: An Update', *Revista da Associacao Medica Brasileira*, 69 (suppl. 1) (2023).

The Last of the Line

Thanks to Dr Alfred Garwood, Anne Karpf, Ruth Evans, Professor Gill Mezey, Dr John Schlabopersky, Dr Anna Motz, Professor Phillipe Sands and Janos Verebes Weisz.

Also to the late Michael Hell and to Mother Angele Arbib of the Abbey of Regina Laudis.

Levi, P. and De Benedetti, L., *Auschwitz Report* (London: Verso Books, 2015).

The Children

Thanks to many colleagues in the Republic of Ireland for their help and advice. And again, to Professors Peter Fonagy and Anthony Bateman for their pioneering work on mentalising, which has been important to so many therapists who see people with complex traumas.

Cicchetti, D., 'Fractures in the Crystal: Developmental Psycho-pathology and the Emergence of Self', *Developmental Review*, 11:3 (1991), pp. 271–87.

———— and Carlson, V. (Eds), *Child Maltreatment: Theory and Research on the Causes and Consequences of Child Abuse and Neglect* (Cambridge: Cambridge University Press, 1989).

Schore, A., *Affect Dysregulation and Disorders of the Self (Norton Series on Interpersonal Neurobiology)* (New York: W. W. Norton & Company, 2003).

The Hostage

Thanks to Dr Vivienne Bailey and Dr Deborah Lee for their insights.

The Target

I am indebted to Nujoji Calvocoressi for his invaluable comments on an earlier draft.

Thanks for their past wisdom and guidance to Professor Nick Abayomi, Dr Anne Aiyegbusi and Professor Aggrey Burke.

Also gratitude to the late Dr Colin Franklin and Gloria Hope Price, dear friends, educators and supporters who made it possible for me to visit their native Barbados often. And again, to Dr Sameer Sarkar for his insights on migration.

Chodorow, N. J., *Femininities, Masculinities, Sexualities: Freud and Beyond* (Lexington: University Press of Kentucky, 1994).

Fanon, F., *Black Skin, White Masks*, trans. Charles Lam Markmann (New York: Grove Press, 2008).

Johnson, T. J., '"Your Silence Will Not Protect You": Using Words and Action in the Fight Against Racism', *Pediatrics*, 149:2 (2022).

The Trainee

Thanks to all the members of the Monday Balint Group (a clinicians peer group). I am grateful for their willingness to discuss difficult cases and talk about being a trainee psychiatrist.

Thanks also to all those who ask me to work with them in supervision – it's a privilege. And thanks to Maja O'Brien, Liza Glenn and Michael Pritchard for helping me to change my mind for good.

Notes

Introduction

1 'With this global pandemic, each and every individual on the surface of the world is affected . . . and that means mass trauma . . . even bigger than what the world experienced after the Second World War.' Tedros Adhanom Ghebreyesus, director general of the World Health Organization, at a virtual press conference, March 2021.

2 Adshead, G., Ferrito, M. and Bose, S., 'Recovery After Homicide: Narrative Shifts in Therapy with Homicide Perpetrators', *Criminal Justice and Behavior*, 42:1 (2014), pp. 70–81; https://doi.org/10.1177/0093854814550030 (original work published 2015). See also Atley, S., Martin, K., Fergus, K. and Goldberg, J., 'I'm Not That Person: A Qualitative Study of Moral Injury in Forensic Psychiatric Patients', *International Journal of Environmental Research and Public Health*, 22 (2025), p. 372.

3 Locke, J., 'An Essay Concerning Human Understanding, 1690', in Dennis, W. (Ed.), *Readings in the History of Psychology* (New York: Appleton-Century-Crofts, 1948), pp. 55–68; https://doi.org/10.1037/11304-008.

4 Hemingway, E., *For Whom the Bell Tolls* (New York: Scribner, 1940).

The POW

1 It would be many years before a civil partnership between same-sex couples was legalised, and nearly twenty-five before gay marriage was permitted in the UK. See https://assets.publishing.service.gov.uk/media/5a750cd2e5274a59fa717007/140423_M_SSC_Act_factsheet__web_version_.pdf.

2 Borges, J. L., 'Funes the Memorious', in *Collected Fictions* (trans. Hurley, A.) (London: Penguin, 1999).

3 https://victoriancollections.net.au/items/
   5a16272d21ea6d034020cbd4. For more background, see Keith
   Wilson's similarly titled book about life in Changi, *You'll Never
   Get Off the Island* (London: Allen & Unwin, 1989).
4 Levi, P., *If This Is a Man* (New York: Orion Press, 1959).
5 Pennebaker, J. W., 'Theories, Therapies, and Taxpayers:
   On the Complexities of the Expressive Writing Paradigm',
   Clinical Psychology: Science and Practice, 11:2 (2004), pp. 138–42.
   Pennebaker, J. W. and Smyth, J. M., *Opening Up by Writing It
   Down: How Expressive Writing Improves Health and Eases Emotional
   Pain* (New York: Guilford Press, 2016).
6 Ranganath, C., *Why We Remember: Unlocking Memory's Power to
   Hold on to What Matters* (New York: Doubleday, 2024).
7 Shakespeare, W., *Henry IV, Part I*, Act II, scene iii;
   https://www.folger.edu/blogs/shakespeare-and-beyond/
   shakespeare-post-traumatic-stress-disorder.
8 Albanese, M., Liotti, M., Cornacchia, L. and Mancini, F.,
   'Nightmare Rescripting: Using Imagery Techniques to Treat
   Sleep Disturbances in Post-traumatic Stress Disorder', *Front
   Psychiatry*, 13 (4 April 2022).
9 Erikson, E., *Identity and the Life Cycle* (New York: W. W. Norton
   & Co., 1980).
10 Calvino, I., *Six Memos for the Next Millennium* (trans. Brock, G.)
   (Boston: Mariner Books, 2016).

The Refugee

1 Murray, L. and Trevarthen, C., 'The Infant's Role in Mother–Infant
   Communications', *Journal of Child Language*, 13:1 (1986), pp. 15–29.
   Murray, L., *The Psychology of Babies: How Relationships Support
   Development from Birth to Two* (London: Hachette UK, 2014).
2 Winnicott, D. W., *The Child, the Family, and the Outside World*
   (Cambridge, MA: Perseus, 1987; originally published 1964).
3 In today's post-Brexit world, this may seem unusual, but at that
   time it was considered quite normal for qualified young people to
   study abroad for periods of time, with programmes like Erasmus
   allowing students to move with relative ease between the UK and
   other countries.

4 Akhtar, S., 'A Third Individuation: Immigration, Identity, and the Psychoanalytic Process', *Journal of the American Psychoanalytic Association*, 43:4 (1995), pp. 1051–84.

5 Summerfield, D., 'War, Exile, Moral Knowledge and the Limits of Psychiatric Understanding: A Clinical Case Study of a Bosnian Refugee in London', *International Journal of Social Psychiatry*, 49:4 (2003), pp. 264–8.

6 Fraiberg, S., Adelson, E. and Shapiro, V., 'Ghosts in the Nursery: A Psychoanalytic Approach to the Problems of Impaired Infant–Mother Relationships 1', in Raphael-Leff, J. (Ed.), *Parent–Infant Psychodynamics*, pp. 87–117 (London: Routledge, 2018). Fonagy, P., Steele, M., Moran, G., Steele, H. and Higgitt, A., 'Measuring the Ghost in the Nursery: An Empirical Study of the Relation Between Parents' Mental Representations of Childhood Experiences and Their Infants' Security of Attachment', *Journal of the American Psychoanalytic Association*, 41:4 (1993), pp. 957–89.

7 Auden, W. H., 'Funeral Blues', in *Another Time* (New York: Random House, 1940).

8 Shakespeare, W., *Macbeth*, Act IV, scene iii.

9 https://www.apa.org/news/podcasts/speaking-of-psychology/grieving-changes-brain.

10 Lewis, C. S., *A Grief Observed* (New York: Harper & Row, 1961).

11 Ibid.

12 Shakespeare, W., *Macbeth*, Act IV, scene iii.

13 Brown, H., 'Tony Blair's Legacy for the National Health Service, Special Report', *The Lancet*, 369:9574 (19 May 2007), pp. 1679–82.

14 Eliot, T. S. 'East Coker', in *Four Quartets* (London: Faber & Faber, 2010).

The Lucky Ones

1 A good contemporary discussion of the complexities of the PTSD diagnosis in a legal context by a person who is a psychiatrist and has lived experience of war-related PTSD comes from Martin Deahl: 'Psychological Trauma and the Law . . . Post-Traumatic Stress Disorder (PTSD) or Post-Traumatic Disorders (PTDs)? Is PTSD Still Fit for Purpose?' *Medicine, Science and the Law*, 65:1 (2025), pp. 3–4.

2 Grey, N., Holmes, E. and Brewin, C. R., 'Peritraumatic Emotional "Hot Spots" in Memory', *Behavioural and Cognitive Psychotherapy*, 29 (2001), pp. 367–72.

3 Horowitz, M., Wilner, N. and Alvarez, W., 'Impact of Event Scale: A Measure of Subjective Stress', *Psychosomatic Medicine*, 41:3 (1979), pp. 209–18.

4 Main, M., 'Introduction to the Special Section on Attachment and Psychopathology: 2. Overview of the Field of Attachment', *Journal of Consulting and Clinical Psychology*, 64:2 (1996), pp. 237–43.

5 The diagnostic criteria for PTSD are still the subject of debate today. See Maercker, A. and Brewin, C. R., 'Controversies in Trauma- and Stress-Related Disorders', *British Journal of Psychiatry*, 226:1 (2025), pp. 7–9. Marx, B. et al., 'The PTSD Criterion – A Debate: A Brief History, Current Status, and Recommendations for Moving Forward, *Journal of Traumatic Stress*, 37:1 (2024), pp. 5–15.

6 Garland, C., *Understanding Trauma: A Psychoanalytical Approach* (part of the Tavistock Clinic series) (London: Routledge, 2002).

7 Magorian, M., *Goodnight Mister Tom* (London: Kestrel, 1981).

8 The diagnostic manual we use in the UK, *The International Classification of Diseases* (published by the World Health Organization), includes a variety of post-natal mental disorders, including post-natal psychosis. Incredibly, at the time of writing, it was not recognised as a distinct syndrome in the *DSM-5*, the gold-standard diagnostic manual for psychiatrists in the US, although American colleagues are campaigning to change this. See this fascinating piece by Cliffel, M. and Hatters Friedman, S., 'Postpartum Psychosis, Two Sides of the Story', *Journal of the American Academy of Psychiatry*, 52 (2024), pp. 486–93.

9 Cho, C., *Inferno: A Memoir of Motherhood and Madness* (London: Bloomsbury, 2020).

The Last of the Line

1 Garwood, A., *Holocaust Trauma and Psychic Deformation: Psychoanalytic Reflections of a Holocaust Survivor* (London: Routledge, 2020).

2 Kandel, E. R., *In Search of Memory: The Emergence of a New Science of Mind* (New York: W. W. Norton & Co., 2005). For further reading, including his astonishing story of resilience and recovery, see Mario Ramberg Capecchi's Wikipedia page: https://en.wikipedia.org/wiki/Mario_Capecchi.

3 Quoted in Neumann, A., 'Past Lives: How Intergenerational Trauma Shapes Mental Health', *Scientific Kenyon: The Neuroscience Edition*, 4:1 (2020), pp. 41–6.

4 Daskalakis, N. P., Xu, C., Bader, H. N., Chatzinakos, C., Weber, P., Makotkine, I., Lehrner, A., Bierer, L. M., Binder, E. B. and Yehuda, R., 'Intergenerational Trauma Is Associated with Expression Alterations in Glucocorticoid- and Immune-Related Genes', *Neuropsychopharmacology*, 46:4 (2021), pp. 763–73.

5 Karpf, A., 'The War After: The Psychology of the Second Generation', *Jewish Quarterly*, 43:2 (1996), pp. 5–11 (later published by Heinemann and Faber as a book).

6 Roethke, T., 'In a Dark Time', in *Collected Poems of Theodore Roethke* (New York: Doubleday, 1961).

7 Frankl, V., *Man's Search for Meaning* (London: Random House, 2004).

8 Niederland, W. G., 'The Survivor Syndrome: Further Observations and Dimensions', *Journal of the American Psychoanalytic Association*, 29:2 (1981), pp. 413–25. See also Levi, P., *The Drowned and the Saved* (New York: Summit Books, 1988).

9 Buechner, F., *Telling Secrets* (San Francisco: Harper, 1991), p. 4. See also his website: https://www.frederickbuechner.com/telling-secrets.

10 A beautiful tribute to Paul Celan and his use of language comes from Neil Arditi, 'Blessing and Rebuke', *Jewish Review of Books*, autumn 2023; https://jewishreviewofbooks.com/jewish-history/14777/blessing-and-rebuke.

11 Kübler-Ross, E. and Kessler, D., *On Grief and Grieving: Finding the Meaning of Grief Through the Five Stages of Loss* (New York: Simon and Schuster, 2005).

12 Popova, M., 'Trial, Triumph, and the Art of the Possible: The Remarkable Story Behind Beethoven's "Ode to Joy"', from *The Marginalian* blog, May 2022; https://www.themarginalian.org/2022/05/17/beethoven-ode-to-joy.

13 Geitel, K., 'Exulting Freedom in Music' (trans. J. and M. Berridge), on Bernstein's website: https://leonardbernstein. com/about/conductor/historic-concerts/berlin-wall-concert-1989.

14 Mithin, S., *The Singing Neanderthals* (Boston: Harvard University Press, 2007).

15 Anwar, Y., 'How Many Emotions Can Music Make You Feel?', *Greater Good Magazine*, 17 January 2020; https://greatergood. berkeley.edu/article/item/how_many_emotions_can_music_ make_you_feel. Cowen, A. S. et al., 'What Music Makes Us Feel: At Least 13 Dimensions Organize Subjective Experiences Associated with Music Across Different Cultures', *Proceedings of the National Academy of Sciences of the United States of America*, 117:4 (2020), pp. 1924–34. BBC Radio 4's classic series *Desert Island Discs* is also a superb example of this.

16 Holmes, R., *Footsteps: Adventures of a Romantic Biographer* (London: Vintage, 1996).

The Children

1 Bradley, J., 'Residential Institutions Redress Board of the Irish Government 2002–2018', *Medico-Legal Journal*, 89:1 (2021), pp. 54–7. The report of the commission has five volumes of evidence: https://childabusecommission.ie/?page_id=241. The IRB was set up under the Residential Institutions Redress Act 2002 to make fair and reasonable awards to persons who, as children, were abused while resident in industrial schools, reformatories and other institutions subject to state regulation or inspection.

2 Some background on the Christian Brothers: Barkham, P., 'The Brothers Grim', *Guardian*, 28 November 2009. Keogh, D., *Edmund Rice and the First Christian Brothers* (Dublin: Four Courts Press, 2008).

3 Cicchetti, D. (Ed.), *Developmental Psychopathology: Risk, Resilience, and Intervention*, vol. 4 (Hoboken: John Wiley & Sons, 2016). Maughan, A. and Cicchetti, D., 'Impact of Child Maltreatment and Interadult Violence on Children's Emotion Regulation Abilities and Socioemotional Adjustment', *Child Development*, 73:5 (2002), pp. 1525–42. Denckla, C. A., Cicchetti, D., Kubzansky,

L. D., Seedat, S., Teicher, M. H., Williams, D. R. and Koenen, K. C., 'Psychological Resilience: An Update on Definitions, a Critical Appraisal, and Research Recommendations', *European Journal of Psychotraumatology*, 11:1 (2020), article no. 1822064. Cohen, P., Crawford, T. N., Johnson, J. G. and Kasen, S., 'The Children in the Community Study of Developmental Course of Personality Disorder', *Journal of Personality Disorders*, 19:5 (2005), pp. 466–86.

4 Original paper: Boyce, W. T. and Ellis, B. J., 'Biological Sensitivity to Context: I. An Evolutionary-Developmental Theory of the Origins and Functions of Stress Reactivity', *Developmental Psychopathology*, 17:2 (2005), pp. 271–301. Subsequent book: Boyce, W. T., *The Orchid and the Dandelion: Why Sensitive People Struggle and How All Can Thrive* (New York: Knopf, 2019).

5 Yeats, W. B., 'The Stolen Child', in *The Wanderings of Oisin and Other Poems to 1895* (Ithaca: Cornell University Press, 1994).

6 The English poet John Keats first wrote about negative capability – the idea of remaining comfortable with uncertainty and doubt, without 'irritable reaching after fact and reason' – in a letter to his brother in 1817, which was later published in Keats, J., *The Complete Poetical Works and Letters of John Keats* (Boston: Houghton, Mifflin and Company, 1899), p. 277.

7 Felitti, V. J. et al., 'Relationship of Childhood Abuse and Household Dysfunction to Many of the Leading Causes of Death in Adults. The Adverse Childhood Experiences (ACE) Study', *American Journal of Preventive Medicine*, 14:4 (1998), pp. 245–58.

8 Madigan, S. et al., 'Adverse Childhood Experiences: A Meta-Analysis of Prevalence and Moderators Among Half a Million Adults in 206 Studies', *World Psychiatry*, 22:3 (2023), pp. 463–71.

9 Russotti, J. et al., 'Child Maltreatment and the Development of Psychopathology: The Role of Developmental Timing and Chronicity', *Child Abuse & Neglect*, 120 (2021), 105215.

10 Fonagy, P., Gergely, G. and Target, M., 'The Parent–Infant Dyad and the Construction of the Subjective Self', *Journal of Child Psychology and Psychiatry*, 48:3–4 (2007), pp. 288–328.

11 Schore, A .N., 'The Effects of Early Relational Trauma on Right Brain Development, Affect Regulation, and Infant Mental

Health', *Infant Mental Health Journal: Official Publication of the World Association for Infant Mental Health*, 22:1–2 (2001), pp. 201–69.

12 Herman, J. L., 'Complex PTSD', in Everly, G. S. and Lating, J. M. (Eds), *Psychotraumatology. The Springer Series on Stress and Coping* (Boston: Springer, 1995).

The Hostage

1 Cummins, I., 'The Impact of Austerity on Mental Health Service Provision: A UK Perspective', *International Journal of Environmental Research and Public Health*, 15:6 (2018), p. 1145; https://pmc.ncbi.nlm.nih.gov/articles/PMC6025145.

2 Beck, J. S. and Fleming, S., 'A Brief History of Aaron T. Beck, MD, and Cognitive Behavior Therapy', *Clinical Psychology in Europe*, 3:2 (2021).

3 Further reading on this: Grounds, A. T., 'Understanding the Effects of Wrongful Imprisonment', *Crime and Justice*, 32 (2005), pp. 1–58.

4 Panchal, N. and Lo, J., 'Exploring the Rise in Mental Health Care Use by Demographics and Insurance Status', KFF research paper, 2024 (www.kff.org).

5 Barker, P., *The Regeneration Trilogy* (London: Penguin, 2013); and because she drew directly from this source, I also recommend Showalter, E., *The Female Malady: Women, Madness, and English Culture, 1830–1980* (New York: Penguin, 1987). See also Baker, J., 'Psychogenic Voice Disorders and Traumatic Stress Experience: A Discussion Paper with Two Case Reports', *Journal of Voice*, 17:3 (2003), pp. 308–18. Note that trauma-related functional aphonias were described in soldiers in both the Boer War and the First World War.

The Target

1 Macpherson, W., *The Stephen Lawrence Inquiry* (London: Stationery Office, 1999).

2 Singh, D., 'NHS Criticised Over Suicide of Mentally Ill Doctor', *British Medical Journal*, 327:7422 (2003), p. 1008.

3 Bogdanova, N. et al., 'Associations Between Sociodemographic Characteristics and Receipt of Professional Diagnosis in Common Mental Disorder: Results from the Adult Psychiatric Morbidity Survey 2014', *Journal of Affective Disorders*, 319 (2022), pp. 112–18.

4 Cooper, C., Spiers, N., Livingston, G., Jenkins, R., Meltzer, H., Brugha, T., McManus, S., Weich, S. and Bebbington, P., 'Ethnic Inequalities in the Use of Health Services for Common Mental Disorders in England', *Social Psychiatry and Psychiatric Epidemiology*, 48 (2013), pp. 685–92. For a contemporary update: Bamrah, J. S., Rodger, S. and Naqvi, H., 'Racial Disparities Influence Access and Outcomes in Talking Therapies', *British Journal of Psychiatry*, 226:4 (2025), pp. 203–5.

5 Collazos, F. and Qureshi, A., 'Cultural Competence in the Mental Health Treatment of Immigrant and Ethnic Minority Clients', *Diversity and Equality in Health and Care*, 2:4 (2005). Sue, D. W. and Torino, G. C., 'Racial-Cultural Competence: Awareness, Knowledge, and Skills', *Racial-Cultural Psychology and Counselling* (2005), p. 1.

6 Office for National Statistics, 'Homicide in England and Wales: Year Ending 2024', https://www.ons.gov.uk/ peoplepopulationandcommunity/crimeandjustice/articles/ homicideinenglandandwales/yearendingmarch2024. And 'Key Facts about Violence', Youth Endowment Fund Report, London; www.youthendowmentfund.org.uk/reports/ key-facts-about-violence/7-in-the-uk-violence-affects-children- from-some-racialised-groups-more-than-others.

7 Owolade, T., *This Is Not America: Why Black Lives in Britain Matter* (London: Atlantic Books, 2023).

8 This study isn't about racial maltreatment, but it is of interest in this context: Danese, A. and Widom, C. S., 'Objective and Subjective Experiences of Child Maltreatment and Their Relationships with Psychopathology', *Nature Human Behaviour*, 4:8 (2020), pp. 811–18.

9 Siddaway, A. P., 'Flashbacks Uniquely Characterise Post-Traumatic Stress Disorder: Distinguishing Flashbacks from Other Psychological Processes and Phenomena', *BJPsych Advances*, published online, 2025, pp. 1–2.

10 Baldwin, J., from a 1961 radio interview for WBAI New York moderated by Nat Hentoff (watch here: https://www.youtube.com/watch?v=jNpitdJSXWY&ab_channel=thepostarchive).

11 Shakespeare, W., *Othello*, Act III, scene iii. And in the present day: Everett, P., *James* (New York: Doubleday, 2024).

The Trainee

1 For description of peer groups/the Balint process, see Fitzgerald, G. and Hunter, M. D., 'Organising and Evaluating a Balint Group for Trainees in Psychiatry', *Psychiatric Bulletin*, 27:11 (2003), pp. 434–6.

2 Bellis, M. A. et al., 'Comparing Relationships Between Single Types of Adverse Childhood Experiences and Health-Related Outcomes: A Combined Primary Data Study of Eight Cross-Sectional Surveys in England and Wales', *British Medical Journal Open*, 13:4 (2023), pp. 1–10.

3 Finkelhor, D., 'Meta-Analysis and Crossnational Comparisons of Sexual Violence Against Children', *JAMA Pediatrics*, 179:3 (2025), pp. 229–30.

4 King James Bible, Psalms 77:4; https://www.kingjamesbibleonline.org/Psalms-77-4.

5 Audre Lorde's talk to the annual Modern Language Association conference in Chicago, 1977. Later published in *The Cancer Journals* (New York: Aunt Lute Books, 1995).

6 For more on situational memory or recalling events within their individual context, see Brewin, C. R., Dalgleish, T. and Joseph, S., 'A Dual Representation Theory of Posttraumatic Stress Disorder', *Psychological Review*, 103:4 (1996), pp. 670–86.

7 Metcalf, O. and Forbes, D., 'Addressing the Double Bind of Women's Anger After Trauma', *JAMA Psychiatry*, 82:5 (2025), pp. 434–6.

8 Beeghly, M. and Cicchetti, D., 'Child Maltreatment, Attachment, and the Self System: Emergence of an Internal State Lexicon in Toddlers at High Social Risk', *Development and Psychopathology*, 6:1 (1994), pp. 5–30. For more on autobiographical competence, see Holmes, J., 'Attachment Theory: A Biological Basis for Psychotherapy?' *British Journal of Psychiatry*, 163:4 (1993), pp. 430–8.

9 Audre Lorde's talk to the annual Modern Language Association conference in Chicago, 1977. Later published in *The Cancer Journals* (New York: Aunt Lute Books, 1995).

Coda

1 Again, a reference to Malcolm's line in *Macbeth*, Act IV, scene iii.